T0338273

Formation Control of Multi-Agent Systems

Wiley Series in Dynamics and Control of Electromechanical Systems

Formation Control of Multi-Agent Systems: A Graph Rigidity Approach

Marcio de Queiroz
Louisiana State University
USA

Xiaoyu Cai
Louisiana State University
USA

Matthew Feemster
United States Naval Academy
USA

Registered Offices
John Wiley & Sons, Inc., 111 River Street, Hoboken, NJ 07030, USA
John Wiley & Sons Ltd, The Atrium, Southern Gate, Chichester, West Sussex, PO19 8SQ, UK

Editorial Office
The Atrium, Southern Gate, Chichester, West Sussex, PO19 8SQ, UK

For details of our global editorial offices, customer services, and more information about Wiley products visit us at www.wiley.com.

Wiley also publishes its books in a variety of electronic formats and by print-on-demand. Some content that appears in standard print versions of this book may not be available in other formats.

Library of Congress Cataloging-in-Publication Data

Names: Queiroz, Marcio S. de, author. | Cai, Xiaoyu, 1987- author. | Feemster, Matthew, author.
Title: Formation control of multi-agent systems : a graph rigidity approach / Professor Marcio de Queiroz (Louisiana State University), Dr. Xiaoyu Cai (Louisiana State University), Dr. Matthew Feemster (United States Naval Academy).
Description: Hoboken, NJ : John Wiley & Sons, Inc., [2019] | Includes bibliographical references and index. |
Identifiers: LCCN 2018040373 (print) | LCCN 2018050691 (ebook) | ISBN 9781118887479 (Adobe PDF) | ISBN 9781118887462 (Epub) | ISBN 9781118887448 (hardcover)
Subjects: LCSH: Multiagent systems. | Formation control (Machine theory) | Graph theory. | Rigidity (Geometry) | Automatic control–Mathematical models. | Robotics–Mathematical models.
Classification: LCC QA76.76.I58 (ebook) | LCC QA76.76.I58 Q84 2019 (print) | DDC 006.3/0285436–dc23
LC record available at https://lccn.loc.gov/2018040373

Cover Design: Wiley
Cover Image: © Kypros/Getty Images, © aapsky/Shutterstock, © Andrei Trentea/Shutterstock

Set in 10/12pt WarnockPro by SPi Global, Chennai, India

Printed and bound by CPI Group (UK) Ltd, Croydon, CR0 4YY

10 9 8 7 6 5 4 3 2 1

To my late father, José
M. de Q.
To my parents, Zhenjie and Chunmei,
and my wife, Bingqing
X.C.
To my parents, Sam and Gay and my family,
Agnes Ann, Sam, Ryn, and Meg
M.F.

Contents

Preface

As the initial hurdles of unmanned robotic platform development have been passed, focus is now being placed on advancing the behavior of these platforms so they perform *coordinated operations in groups* with and without human supervision. Over the past several years, a considerable amount of work has been conducted in this area under various names: multi-agent systems, networked systems, cooperative control, and swarming. Research has evolved from fundamental studies of biological swarms in nature to the development and application of systems theoretical tools for modeling such behaviors to, more recently, the synthesis and experimental validation of engineered multi-agent systems.

The premise behind engineering multi-agent systems is that cooperation among group members can lead to the execution of complex functions that are otherwise not possible. Engineering multi-agent systems have the potential to impact a variety of military, civilian, and commercial applications that involve some of form situational awareness. Examples include patrolling, monitoring, surveying, scouting, and element tracking over large geographical areas with unmanned robotic vehicles or mobile sensor networks.

Decentralization is a key characteristic of biological and engineered multi-agent systems since it provides adaptability and robustness to the system operation. Several coordination-type problems have been studied within the robotics, systems, and control research communities that involve some level of distributed operation. Graph theory plays an important role in modeling the decentralization and interaction among the multiple agents needed to achieve the common goal. Our interest in this book is in the class of coordination problems known as *formation control* and in the use of *rigid graph theory* as a solution tool. Specifically, the goal of the book is to provide the first comprehensive and unified treatment of the subject of graph rigidity-based formation control of multi-agent systems. The presentation is mostly based on the authors' own work and perspectives.

The book begins with an introduction to rigid graph theory for readers not familiar with the subject. The heart of the book is divided into three parts according to the model of the agents' equations of motion: the single-integrator model, the double-integrator model, and the robotic vehicle model. For each model, three types of formation problems are studied: formation acquisition, formation maneuvering, and target interception. All formation control results in the book are supported by computer simulations, while most are demonstrated experimentally using unmanned ground vehicles. The book is organized such that the material is presented in ascending level of difficulty, building upon previous sections and chapters.

The book is intended for researchers and graduate students in the areas of robotics, systems, and control who are interested in the topic of multi-agent systems. We assume readers have a graduate-level knowledge of linear algebra, matrix theory, control systems, and nonlinear systems, especially Lyapunov stability theory.

We would like to acknowledge and express our gratitude to Pengpeng Zhang and Milad Khaledyan for their assistance with some of the theoretical results and computer simulations presented in the book, and to Dr. Bingqing Wu for her assistance with the creation of Figures 1.3 and 1.5. We would also like to thank Eric Willner and Jemima Kingsly at Wiley for giving us the opportunity to publish this work and for their patience while we completed it.

Finally, we acknowledge the following entities for allowing us to reproduce their pictures:

- Weaver ants making an emergency bridge between two plants by Rose Thumboor (see Figure 1.1). Retrieved from commons.wikimedia.org/ wiki/File:Weaver_Ants_-_Oecophylla_smaragdina.jpg. Used under Creative Commons Attribution-Share Alike 4.0 International license (creative commons.org/licenses/by-sa/4.0/deed.en).
- School of convict surgeonfish (*Acanthurus triostegus*) by Thomas Shahan (see Figure 1.1). Retrieved from www.flickr.com/photos/49580580 @N02/ 14280168344/. Used under Creative Commons Attribution 2.0 Generic license (creativecommons.org/licenses/by/2.0/).
- xBee module (see Figure 5.3). Retrieved from www.sparkfun.com/products/ 8665?. Used under Creative Commons Attribution 2.0 Generic license (creativecommons.org/licenses/by/2.0/).

March 2018

Baton Rouge, LA, USA
Marcio de Queiroz

Exton, PA, USA
Xiaoyu Cai

Annapolis, MD, USA
Matthew Feemster

About the Companion Website

This book is accompanied by a companion website:

www.wiley.com/go/dequeiroz/formation_control

The website material consists of MATLAB files for most of the computer simulations

Scan this QR code to visit the companion website.

1

Introduction

"The whole is more than the sum of its parts."

Aristotle

1.1 Motivation

This book is devoted to *multi-agent systems*. Since this term has different meanings within different research communities, we deem it necessary to precisely define the meaning used here. In this book, a multi-agent system refers to a network of interacting, mobile, *physical* entities that *collectively* perform a complex task beyond their individual capabilities.

Nature is replete with biological systems that fit this definition: a flock of birds, a school of fish, and a colony of insects (see Figure 1.1), to name a few. The behavior of such biological swarms is decentralized since each biological agent does not have access to global knowledge or supervision, but uses its own local sensing, decision, and control mechanisms.

Ants are a model example of a biological multi-agent system. Ant colonies share the common goals of surviving, growing, and reproducing. Their sense of community is so strong that they behave like a single "superorganism" that can solve difficult problems by processing information as a collection [1]. This collective behavior facilitates food gathering, defending nests against enemies, and building intricate structures with tunnels, chambers, and ventilation systems. Ants accomplish such feats without a supervisor telling them what to do. Rather, ant workers perform tasks based on personal aptitudes, communications with colony mates, and cues from the environment. Interactions with other ants and the environment occur via chemicals, which they sense with their antennae [1].

Nature is inspiring humans to *engineer* multi-agent systems that mimic this distributed, coordinated behavior. The agents in such engineering systems are not living beings, but machines such as robots, vehicles, and/or mobile

Formation Control of Multi-Agent Systems: A Graph Rigidity Approach, First Edition.
Marcio de Queiroz, Xiaoyu Cai, and Matthew Feemster.
© 2019 John Wiley & Sons Ltd. Published 2019 by John Wiley & Sons Ltd.
Companion website: www.wiley.com/go/dequeiroz/formation_control

Figure 1.1 Examples of collective behavior in nature: a flock of birds (top left), a school of fish (top right), and a swarm of ants building a bridge (bottom).

sensors (see Figure 1.2). Recent advances in sensor technology, embedded systems, communication systems, and power storage now make it feasible to deploy such swarms of cooperating agents for various civilian and military applications. For instance, a group of autonomous (ground, underwater, water surface, or air) vehicles could be deployed in large disaster areas to perform search, mapping, surveillance, or environmental monitoring and clean up without putting first responders in harm's way. Some recent examples of such situations are Hurricane Katrina in 2005, the BP oil spill in the Gulf of Mexico in 2010, and the Fukushima nuclear disaster in 2011. Another application is a military mission where a group of unmanned air vehicles surround and intercept an intruding or evading aircraft or enemy combatants. Yet another potential application is a team of vehicles cooperatively transporting an object too large and/or heavy for a single vehicle to transport.

One may wonder why a multi-agent system should be used instead of a single "large agent". There are several advantages to doing so: more efficient and complex task execution, robustness when one or more agents fail, scalability,

Figure 1.2 Examples of engineering multi-agent systems.

versatility, adaptability, and lower cost [2]. For example, multiple agents could position themselves relative to each other to create a virtual, large-scale antenna with higher sensitivity to acoustic signals than would be possible with a single antenna. If one of the agents malfunctions, the remaining ones would reconfigure to keep the antenna operational, whereas the stand-alone antenna would be a single point of failure. Malfunctions of an agent are also less likely than in a single system because they are usually much simpler hardware- and software-wise. This simplicity, along with larger quantities, also leads to mass production at a low cost.

On the other hand, multi-agent systems introduce a host of unique challenges, including coordination and cooperation schemes, distribution of information and subtasks, negotiation between team and individual goals, communication protocols, sensing, and collision avoidance. These challenges are exacerbated by the fact that often the task is to be completed with limited computational, communication, and sensing resources. A key design decision is between a centralized coordination scheme and a decentralized/distributed one. In a centralized scheme, each agent has access to measurement and/or control information from a master entity, such as a central processing unit or a global positioning system (GPS). Therefore, centralized schemes have a single point of failure like a single "large agent". They also do not scale well with the number of agents because the processing overhead and number of communication links become prohibitive.

The multi-agent literature has numerous references to decentralized and distributed schemes with the two terms used interchangeably. Unfortunately, there does not seem to be a precise definition for either concept, with different authors using different definitions. In order to avoid misunderstandings, it is important that we define what we mean by decentralized/distributed in this book.

Definition 1.1   A decentralized (or distributed) multi-agent coordination scheme is one where any information—cognitive, sensory, computational, control, etc.—is acquired **locally** by each agent via **onboard** hardware and software. This includes sensors, communication links, memory, and processors. That is, only the agents themselves are responsible for acquiring information and sharing it as necessary without external aid. Information can be input to an agent by onboard sensors or wireless communication with other agents, or pre-programmed into the agent.

Sometimes it is debatable if a coordination scheme can classified as centralized or decentralized. For example, in a leader–follower scheme, the leader agent can be viewed as the master entity despite being a local component of the multi-agent system. However, the coordination scheme can be designed such that if the leader malfunctions, another agent takes up this role with minimal disruption to the assignment. According to Definition 1.1, a leader–follower scheme where the leader is an integral part of the multi-agent system is deemed decentralized in this book.

A number of coordination and cooperation problems for multi-agent systems are described in the literature: aggregation, consensus, agreement, rendezvous, synchronization, social foraging, flocking, coverage, scheduling, and formation [2–4]. Most of these problems are similar in that the purpose is to drive the multiple agents to some common state (position, velocity, frequency, arrival time, temperature, voltage, etc.), which in general may not be related to their motion. This book is devoted strictly to the class of *formation* problems. Specifically, our focus is on three related problems with increasing levels of complexity: *formation acquisition* (where agents are required to form and maintain a pre-defined geometric configuration in space), *formation maneuvering* (where agents are required to simultaneously acquire a formation and move as a unit following a pre-defined trajectory), and *target interception* (where agents intercept and surround a moving target with a pre-defined formation). Note that formation acquisition is a pre-condition for formation maneuvering and target interception. An example of an application where formation control is necessary is the use of a fleet of unmanned air vehicles (UAVs) to create an aerial image of a large area with high spatial resolution. The UAVs need to be properly positioned relative to each other such that each UAV image can be stitched together, with no gaps, to form the overall area map. Another example is maintaining the optimal placement of a network of mobile sensors during static and dynamic target tracking applications such that the collective sensing performance is improved [5]. The classification of formation problems used in this book may have elements of and partially overlap with some of the coordination/cooperation problems listed above. For instance, formation maneuvering is related to flocking, where agents have to move cohesively without colliding with each other.

Often in formation control the main concern is convergence to the desired spatial configuration irrespective of its exact global position in space. That is, the formation is to be acquired up to rotation and translation of the whole set of agent positions. This means that only the *relative positions* of agents need to be known by the control algorithm. Formation control problems are relatively straightforward to solve when the agents' absolute coordinates (i.e., with respect to an Earth-fixed coordinate frame) are available via a central planner. From this information, the relative positions can be readily calculated. However, a GPS, which is typically used in such cases, can lose accuracy when the line of sight between the GPS receiver and satellite is obstructed (e.g., urban areas, dense vegetation, underwater). Therefore, one would like to use a decentralized formation scheme where information is obtained from a suite of onboard sensors (see Figure 1.3). A number of commercially available sensors can be used for this purpose: inertial-type navigation system, laser range finder, sonar, radar, infrared sensor, camera, and compass.

Accurate control of the agents' relative positions is critical for solving formation problems. This, in turn, is strongly dependent on the model of the agents' motion used to design the formation controller. A common approach is to design a "high-level" control law by assuming that the agent motion is governed by the single-integrator model. In this model, each agent is treated as a *kinematic* point where the state is position and the control input is velocity. Another common but lower level approach is to use the double-integrator model. Here, each agent is treated as a holonomic point mass where the states are position and velocity, and the control input is acceleration. That is, the double-integrator model is a *dynamic* model, albeit simplified, of the agent motion. Neither of these simplified approaches can be directly implemented on an actual multi-agent system since they do not provide actuator-level (i.e., force/torque) inputs. At best, they can be embedded as outer control loops in a nonmodel-based, actuator-level control system that neglects the agent dynamics. For applications where high performance is expected, this approach

Figure 1.3 Centralized (left) and decentralized (right) formation control.

may not yield the necessary control accuracy due to improper compensation of the dynamic effects influencing the agent motion (e.g., inertial, frictional, gravitational, nonholonomic, actuator, etc.). In such cases, the explicit compensation of these effects can only be accomplished by considering the full agent dynamics in the control design.

Graph theory is a natural tool for describing the multi-agent formation shape as well as the inter-agent sensing, communication, and control topology in the decentralized case. This book is based on an important subset of this theory—*rigid graph theory*—since it naturally ensures that the inter-agent distance constraints of the desired formation are enforced through the graph rigidity. This implicitly ensures that collisions between agents are avoided while acquiring the formation. We use the concept of graph rigidity as an abstraction of the rigidity of physical structures. In our case, the "vertices" of the "structure" are the agents and the "bars" connecting the vertices are the inter-agent distance constraints imposed by the desired formation. Within this framework, it is convenient to treat each agent as a point and model their motion with the single-integrator equation. As a result, most graph rigidity-based formation controllers utilize the single-integrator model. In this book we will go beyond this approach and introduce results that are based on the double-integrator model and, subsequently, the full dynamic model.

Independent of the model used to describe the agent motion, the formation control laws will primarily stabilize the inter-agent distance dynamics to desired distances. Despite controlling the inter-agent distances, the control laws will depend on the distance *and* angle between agents (i.e., the relative position vector). This is a case of over-sensing [6, 7], but it is a necessary feature of distance-based formation control laws to the best of our knowledge. We will, however, minimize the number of relative positions that need to be measured by making use of a special property of rigid graphs called *minimal rigidity*.

In addition to concepts from rigid graph theory, our control designs will make use of nonlinear stability theory [8] and the integrator backstepping control technique [9]. The former is needed since the inter-agent distance dynamics are nonlinear. As a result, the stability analyses will require an interplay of rigid graph theory and nonlinear analysis methods. The latter will allow us to seamlessly incorporate the single-integrator control designs into the double-integrator and holonomic dynamics designs.

1.2 Notation

In this section, we compile some mathematical notations used throughout the book. The material is provided primarily for reference purposes and is likely familiar to most readers. Some relevant supporting definitions and results from matrix theory, linear algebra, signals, and system theory are provided in Appendices A to C.

- $x \in \mathbb{R}^n$ or $x = [x_1, \ldots, x_n]$ denotes an $n \times 1$ (column) vector.
- $x = [x_1, \ldots, x_n]$ where $x_i \in \mathbb{R}^m$ denotes the stacked $mn \times 1$ vector.
- $\|x\|$ denotes the Euclidean norm or 2-norm of the vector x (we omit the subscript 2 to simplify the notation); $\|x\|_1$ denotes the vector 1-norm.
- For points $\zeta, x \in \mathbb{R}^n$ and set $\mathcal{M}$,

$$\mathrm{dist}(\zeta, \mathcal{M}) := \inf_{x \in \mathcal{M}} \|\zeta - x\|,$$

  i.e., the smallest distance from point ζ to any point in $\mathcal{M}$.
- The convex hull of points x_i, $i = 1, \ldots, n$ in Euclidean space is denoted by $\mathrm{conv}\{x_1, \ldots, x_n\}$ and represents the smallest convex set that contains all points.
- $\mathbf{1}_n$ is the $n \times 1$ vector of ones.
- I_n is the $n \times n$ identity matrix.
- 0 can mean the scalar zero or a vector of all zeros depending on the situation.
- $\lambda_{\min}(\cdot)$ and $\lambda_{\max}(\cdot)$ denote the minimum and maximum eigenvalues of a matrix, respectively.
- $\mathrm{diag}(a_1, \ldots, a_n)$ where $a_i \in \mathbb{R}$ denotes the $n \times n$ diagonal matrix with diagonal elements a_i.
- $\mathrm{diag}(A_1, \ldots, A_n)$ where $A_i \in \mathbb{R}^{m \times m}$ denotes the $mn \times mn$ block diagonal matrix.

1.3 Graph Theory

The control algorithms in this book will rely on some basic concepts of rigid graph theory in two and three dimensions. Below we provide an introduction to these concepts. Henceforth, the superscript m in $\mathbb{R}^m$, which denotes the Euclidean space of the graph, will be either 2 or 3.

1.3.1 Graph

An undirected graph G is a pair (V, E) where $V = \{1, 2, \ldots, n\}$ is the set of vertices and $E \subset V \times V$ is the set of undirected edges such that if vertex pair $(i, j) \in E$ then so is (j, i) (i.e., the vertex pair is not ordered and is counted only once). Since we will only be dealing with undirected graphs in this book, we will drop the term "undirected" from here on when referring to a graph, with the understanding that the concepts in this chapter are introduced in the context of undirected graphs.

For any graph, the number of edges l belongs to the set

$$l \in \{1, \ldots, n(n-1)/2\}. \tag{1.1}$$

We will use the notation $G = (V, E)$ to denote that graph G has vertex set V and edge set E. We only consider graphs where $n > m$ to discount uninteresting, special cases. The set of neighbors of vertex i will be represented by

$$\mathcal{N}_i(E) = \{j \in V \mid (i, j) \in E\}, \tag{1.2}$$

i.e., the set of all vertices that are connected to vertex i with an edge.

A graph G is said to be *complete* if every pair of distinct vertices is connected by an edge, i.e., $l = n(n-1)/2$. A complete graph with n vertices is symbolized by K_n. Figure 1.4 shows an example of the complete graph K_5.

A *path* is a trail that goes from an origin vertex to a destination vertex by traversing edges of the graph. Two vertices i and j are said to be *connected* if there exists a path between these vertices. A graph is connected if there is a path between every pair of vertices of G.

The *adjacency matrix* $\mathcal{A} = [a_{ij}] \in \mathbb{R}^{n \times n}$ of a graph $G = (V, E)$ is defined such that

$$a_{ij} = \begin{cases} 1, & \text{if } (i, j) \in E \\ 0, & \text{otherwise} \end{cases}, \quad a_{ij} = a_{ji}, \ i \neq j, \quad \text{and} \quad a_{ii} = 0. \tag{1.3}$$

The *Laplacian matrix* $\mathcal{L} = [\ell_{ij}] \in \mathbb{R}^{n \times n}$ is defined as

$$\ell_{ii} = \sum_{j=1}^{n} a_{ij} \quad \text{and} \quad \ell_{ij} = -a_{ij}, i \neq j. \tag{1.4}$$

Note that the Laplacian matrix is symmetric and satisfies

$$\sum_{j=1}^{n} \ell_{ij} = 0, \quad i = 1, \dots, n. \tag{1.5}$$

The Laplacian matrix has some other interesting and useful properties. If $\lambda_1 \leq \cdots \leq \lambda_n$ denote the n eigenvalues of $\mathcal{L}$, then $\lambda_1 = 0$ and $\lambda_2 \geq 0$. That is,

Figure 1.4 The complete graph K_5.

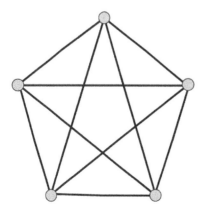

$\mathcal{L}$ is a positive semi-definite matrix. If G is connected, then $\lambda_2 > 0$ (i.e., $\mathcal{L}$ has a single zero eigenvalue). In this case, the eigenvector associated with λ_1 is $\mathbf{1}_n$ such that

$$\mathcal{L}\mathbf{1}_n = 0. \tag{1.6}$$

This implies that

$$\mathcal{L}x = 0$$

if and only if $x \in \mathbb{R}^n$ with $x_i = x_j$, $\forall i, j$.

Lemma 1.1 *[10]* Let $B = \operatorname{diag}(b_1, \dots, b_n)$ be such that $b_i = 1$ or 0, $i = 1, \dots, n$ with at least one nonzero entry, then the matrix

$$M = \mathcal{L} + B$$

is symmetric and positive definite.

1.3.2 Framework

A framework is simply a realization of a graph at given points in Euclidean space. Specifically, if $p_i \in \mathbb{R}^m$ is the coordinate of vertex i with respect to some fixed coordinate frame, then a framework F is a pair (G, p) where $p = [p_1, \dots, p_n] \in \mathbb{R}^{nm}$. We will use the notation $F = (G, p)$ to denote that framework F is composed of graph G with coordinates p. The importance of a framework is that it can model a physical structure. That is, consider the so-called "bar-and-joint" framework, which is a structure made of rigid bars joined at their ends by universal joints.[1] A bar-and-joint framework can be used to describe a wide range of static and dynamic structures, such as bridges, mechanical linkages, and biological molecules (see Figure 1.5).

We are often interested in knowing the length of the edges in a framework. Based on an arbitrary ordering of the edges in E, the edge function $\phi : \mathbb{R}^{nm} \to \mathbb{R}^l$ for a framework $F = (G, p)$ where $G = (V, E)$ is given by

$$\phi(p) = [\dots, \|p_i - p_j\|^2, \dots], \quad (i, j) \in E. \tag{1.7}$$

The kth component of (1.7), $\|p_i - p_j\|^2$, corresponds to the kth edge of E connecting vertices i and j.

Example 1.1 Figure 1.6a shows a graph $G = (V, E)$ in $\mathbb{R}^2$ with $n = 5$ and $l = 6$. By numbering the vertices and edges arbitrarily, we have

$$V = \{1, 2, 3, 4, 5\} \quad \text{and}$$
$$E = \{(1, 2), (2, 3), (3, 4), (1, 4), (2, 4), (4, 5)\}. \tag{1.8}$$

---

1 A universal joint is one where, if one of the adjacent bars is fixed, the other bar can rotate in every direction.

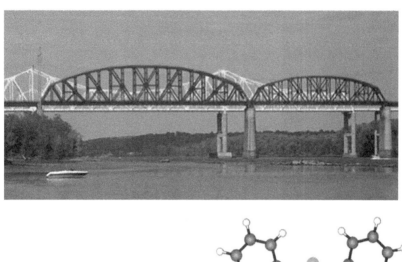

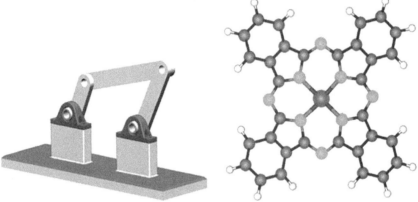

Figure 1.5 Examples of bar-and-joint frameworks, although the joints here are not necessarily universal.

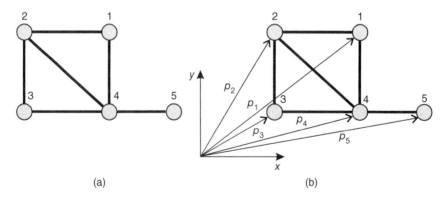

(a)   (b)

Figure 1.6 Example of (a) a graph and (b) a corresponding framework.

Note that the maximum possible number of edges when $n = 5$ is $l = 10$, which is obtained by adding the following edges to E: $(1, 3)$, $(1, 5)$, $(2, 5)$, and $(3, 5)$. The set of neighbors of, for example, vertices 2 and 5 are $\mathcal{N}_2(E) = \{1, 3, 4\}$ and $\mathcal{N}_5(F) = \{4\}$, respectively. Figure 1.6b shows a framework F associated with G by assigning coordinates to each vertex, i.e., $F = (G, p)$ where $p = [p_1, p_2, p_3, p_4, p_5]$, $p_i = [x_i, y_i]$, and G was defined by (1.8). According to (1.7), the edge function for F is given by

$$\phi(p) = [\|p_1 - p_2\|^2, \|p_2 - p_3\|^2, \|p_3 - p_4\|^2, \|p_1 - p_4\|^2,$$
$$\|p_2 - p_4\|^2, \|p_4 - p_5\|^2]. \tag{1.9}$$

1.3.3 Rigid Graphs

Given a bar-and-joint framework, a fundamental question is whether it is *rigid* or not. By rigidity, we mean the non-deformation of the structure. The concept of *rigid* graphs is central to this book, and is an abstraction of the rigidity of civil and mechanical structures. The first reference to graph rigidity dates back to a mathematical problem studied by Leonhard Euler in 1766 [12]. Our goal here is to have a simple means of predicting rigidity. To formalize the concept of rigidity, we need to first introduce the following definitions.

Definition 1.2 *[11]* A rigid body is said to be in translation if all points forming the body move along parallel paths (straight or curvilinear). A rigid body is said to be in rotation if all points move in parallel planes along circles centered on a same axis that intersects the body.

If the axis of rotation does not intersect the rigid body, the motion is usually called a revolution or orbit. This type of motion (in fact, any type of rigid body motion) can be decomposed into a translation superimposed on a rotation. That is, the above definitions allow one to *decouple* pure translation from pure rotation. Recall from rigid body kinematics that given body-fixed points p and O (see Figure 1.7), the following relationship holds

$$\dot{p} = \dot{R} + \omega \times r \tag{1.10}$$

where $\omega \in \mathbb{R}^3$ denotes the angular velocity of the rigid body about an arbitrary axis $\hat{R}$ passing through point O and $\{x, y, z\}$ is an inertial coordinate frame.

Definition 1.3 *[13]* Two frameworks (G, p) and $(G, \hat{p})$ with $G = (V, E)$ are:

- Equivalent if $\|p_i - p_j\| = \|\hat{p}_i - \hat{p}_j\|$ for all $(i, j) \in E$, i.e., the edge function (1.7) is the same;

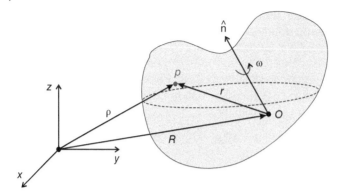

Figure 1.7 Rigid body kinematics.

- Congruent if $\|p_i - p_j\| = \|\hat{p}_i - \hat{p}_j\|$ for all $i, j \in V$ (with $i \neq j$),[2] i.e., all distances between vertices are the same.

Note that congruency implies equivalency, but the reverse is not necessarily true.

Definition 1.4 *[14]* An isometry of $\mathbb{R}^m$ is a bijective map $T : \mathbb{R}^m \to \mathbb{R}^m$ such that

$$\|T(x) - T(y)\| = \|x - y\|, \quad \forall x, y \in \mathbb{R}^m. \tag{1.11}$$

Note that T accounts for rotation and/or translation of the vector $x - y$.

Two frameworks (G, p) and $(G, \hat{p})$ are said to be *isomorphic* in $\mathbb{R}^m$ if they are related by an isometry in $\mathbb{R}^m$. It is not difficult to see that isomorphic frameworks are congruent. We denote the set of all frameworks that are isomorphic to F by $\text{Iso}(F)$.

Example 1.2 Consider the framework F in $\mathbb{R}^2$ shown in Figure 1.8, which is an equilateral triangle with sides of length a. Frameworks F_1 and F_2 are isomorphic to F since they can be obtained by translating and/or rotating F.

Definition 1.5 *[14]* A motion of a framework $F = (G, p)$ with $G = (V, E)$ is a continuous family of equivalent frameworks $F(t)$ for $t \in [0, 1]$ where $F(0) = F$. That is, each point p_i, $i \in V$ moves along a continuous trajectory $p_i(t)$ while preserving the distances between points connected by an edge.

---

2 Henceforth, whenever the notation $i.j \in V$ is used, it should be understood that $i \neq j$. Note that the number of such (i, j) pairs is always $n(n - 1)/2$ according to (1.1).

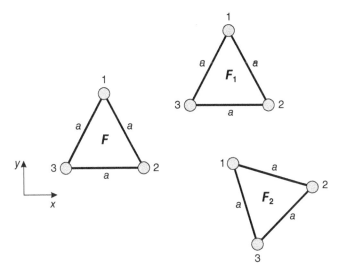

Figure 1.8 Isomorphic frameworks.

This leads us to the following definition of rigidity.

Definition 1.6 *[14, 15]*   A framework $F = (G, p)$ is rigid in $\mathbb{R}^m$ if all of its motions satisfy $p_i(t) = T(p_i)$, $\forall i \in V$, and $\forall t \in [0, 1]$, i.e., the family of frameworks $F(t)$ is isomorphic. On the other hand, the framework is flexible in $\mathbb{R}^m$ if and only if it is possible to continuously move its vertices to form an equivalent but non-congruent framework.

Example 1.3   The bar-and-joint framework in Figure 1.9a is flexible in $\mathbb{R}^2$ because the motion of vertices 1 and 2 leads to a framework that, although equivalent, is non-congruent to the original one. In particular, the distances between vertices 1 and 3 and vertices 2 and 4 have been altered. This deformation can be prevented by adding an edge (bar) between vertices 1 and 3, for example. Therefore, the framework in Figure 1.9b is now rigid and its motions are restricted to rigid body ones. Notice that the bridge in Figure 1.5 is designed to be a rigid framework (at least one would hope), while the four-bar

Figure 1.9 Examples of (a) flexible and (b) rigid bar-and-joint frameworks.

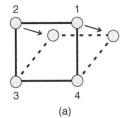

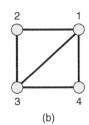

(a)   (b)

mechanism is not. Finally, the framework in Figure 1.6 is flexible since edge (4,5) can be rotated about vertex 4, changing the distance between vertices 1 and 5, for example.

1.3.4 Infinitesimal Rigidity

Although our physical intuition can indicate if certain frameworks are rigid or not, in general it is difficult to determine rigidity from Definition 1.6. Therefore, we will consider a related notion of rigidity called *infinitesimal rigidity*, which can be easily verified via a matrix rank. In general, rigidity does not imply infinitesimal rigidity, but infinitesimal rigidity implies rigidity. Frameworks that are rigid but not infinitesimally rigid usually have collinear or parallel edges. For example, the framework in Figure 1.10 is rigid but not infinitesimally rigid since it has three parallel edges. The reader is referred to [14, 16] for other examples.

For so-called *generic* frameworks, rigidity is equivalent to infinitesimal rigidity [15].[3] When a framework $F = (G, p)$ is generic, then *almost all* of its realizations have the same rigidity property under small perturbations on p [16]. As a result, generic rigidity is a property only of the underlying graph G, i.e., it is independent of almost all p. The term "almost all" is used to exclude certain degenerate configurations, such as frameworks that lie in a hyperplane.[4] A detailed study of generic frameworks can be found in [17].

Example 1.4 The planar framework in Figure 1.11 is nongeneric in $\mathbb{R}^2$ since vertices 1, 2, 3, and 4 are collinear. The planar frameworks in Figure 1.9 are generic in $\mathbb{R}^2$, but nongeneric in $\mathbb{R}^3$ since all vertices are in a plane.

In infinitesimal rigidity, instead of preserving distances during a continuous deformation, we require *first-order* preservation of distances during an *infinitesimal* motion. Therefore, infinitesimal rigidity (also called first-order

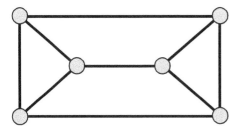

Figure 1.10 Example of a rigid framework that is not infinitesimally rigid.

---

3 It has been noted in the literature (see [15, 18]) how these nuanced notions of rigidity often cause confusion, with definitions varying from author to author.

4 A hyperplane is a subspace of one dimension less than its Euclidean space. For example, a line is a hyperplane in $\mathbb{R}^2$ and a plane is a hyperplane in $\mathbb{R}^3$.

Figure 1.11 A nongeneric framework in $\mathbb{R}^2$.

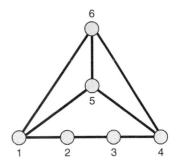

rigidity) is a linear approximation to rigidity. To be more precise, we first use Definition 1.5 to write

$$\|p_i(t) - p_j(t)\| = \|p_i - p_j\|, \quad (i, j) \in E. \tag{1.12}$$

Note that the right-hand side of (1.12) is constant by definition. Assuming the trajectories $p_i(t)$, $i \in V$ are differentiable on $t \in [0, 1]$, we square both sides of (1.12) and differentiate with respect to time to obtain

$$\frac{d}{dt}\|p_i(t) - p_j(t)\|^2 = 2(p_i(t) - p_j(t))^\top (\dot{p}_i(t) - \dot{p}_j(t)) = 0, \quad (i, j) \in E, \tag{1.13}$$

where $\dot{p}_i$ denotes the time derivative of p_i. Rather than require that (1.13) be satisfied for all $t \in [0, 1]$, we impose the condition that it only need to hold at $t = 0$:

$$(p_i - p_j)^\top (v_i - v_j) = 0, \quad (i, j) \in E, \tag{1.14}$$

where $v_i := \dot{p}_i(0)$ and, from Definition 1.5, $p_i(0) = p_i$. The above equation leads to a system of l linear equations with the nm unknowns being the velocities v_i. When instantaneous velocities v_i that satisfy (1.14) exist, we say the framework has infinitesimal motion.

When analyzing infinitesimal rigidity, the *rigidity matrix* of a framework comes in handy. The rigidity matrix $R : \mathbb{R}^{nm} \to \mathbb{R}^{l \times nm}$ is defined as

$$R(p) = \frac{1}{2}\frac{\partial \phi(p)}{\partial p} \tag{1.15}$$

where $\phi(p)$ was given in (1.7). Note that the rigidity matrix has a row for each edge and m columns for each vertex. That is, for the kth edge of E connecting vertices i and j, the kth row of R is

$$[0 \ \ldots \ 0 \ (p_i - p_j)^\top \ 0 \ \ldots \ 0 \ (p_j - p_i)^\top \ 0 \ \ldots \ 0] \tag{1.16}$$

where $(p_i - p_j)^\top$ is in the columns for vertex i, $(p_j - p_i)^\top$ is in the columns for vertex j, and all other elements are zero.

Example 1.5 The rigidity matrix of the framework with edge function (1.9) is given by

$$
R(p) = \begin{bmatrix}
(p_1 - p_2)^{\mathsf{T}} & (p_2 - p_1)^{\mathsf{T}} & 0 & 0 & 0 \\
0 & (p_2 - p_3)^{\mathsf{T}} & (p_3 - p_2)^{\mathsf{T}} & 0 & 0 \\
0 & 0 & (p_3 - p_4)^{\mathsf{T}} & (p_4 - p_3)^{\mathsf{T}} & 0 \\
(p_1 - p_4)^{\mathsf{T}} & 0 & 0 & (p_4 - p_1)^{\mathsf{T}} & 0 \\
0 & (p_2 - p_4)^{\mathsf{T}} & 0 & (p_4 - p_2)^{\mathsf{T}} & 0 \\
0 & 0 & 0 & (p_4 - p_5)^{\mathsf{T}} & (p_5 - p_4)^{\mathsf{T}}
\end{bmatrix}.
$$

$$(1.17)$$

Using (1.15), we can conveniently rewrite (1.14) as

$$
R(p)v = 0 \tag{1.18}
$$

where $v = [v_1, \ldots, v_n] \in \mathbb{R}^{nm}$. Therefore, infinitesimal motion exists when (1.18) has a nontrivial (nonzero) solution v. We would like to know when this is the case.

Note that if a framework undergoes any rigid body motion, then according to (1.10) vertex i has velocity

$$
v_i = v^* + \omega \times p_i \tag{1.19}
$$

where $v^* \in \mathbb{R}^3$ is the translational velocity and $\omega \in \mathbb{R}^3$ is the angular velocity.[5] It is not difficult to check that (1.18) holds in this case since, from (1.16), we have that

$$
\begin{aligned}
(p_i - p_j)^{\mathsf{T}} v_i + (p_j - p_i)^{\mathsf{T}} v_j &= (p_i - p_j)^{\mathsf{T}} (v^* + \omega \times p_i) \\
&\quad + (p_j - p_i)^{\mathsf{T}} (v^* + \omega \times p_j) \\
&= (p_i - p_j) \cdot (\omega \times p_i) + (p_j - p_i) \cdot (\omega \times p_j) \\
&= p_i \cdot \omega \times p_i - p_j \cdot \omega \times p_i + p_j \cdot \omega \times p_j \\
&\quad - p_i \cdot \omega \times p_j \\
&= -p_j \times \omega \cdot p_i - p_i \cdot \omega \times p_j = 0
\end{aligned} \tag{1.20}
$$

upon use of the properties of vector products. In other words, rigid body motions produce infinitesimal motions. This leads us to the following definition.

Definition 1.7 *[16]* A framework is infinitesimally rigid if the only solutions to (1.18) arise from rigid body motions. Otherwise, it is infinitesimally flexible.

---

5 Notice that the vectors are defined here with dimension 3 irrespective of the Euclidean space of the framework. If the framework is planar, then $v^* = [v_x^*, v_y^*, 0]$, $\omega = [0, 0, \omega_z]$, and $p_i = [p_{ix}, p_{iy}, 0]$.

A useful structural property of the rigidity matrix, which follows from (1.16) and will be exploited in some of the control designs of this book, is the following.

Property 1.1 Given any vector $x \in \mathbb{R}^m$,

$$R(p)(\mathbf{1}_n \otimes x) = 0.$$

We can use Definition 1.7 to determine if a framework is infinitesimally rigid or flexible by attempting to assign a velocity vector to each vertex such that (1.18) holds. For example, consider the framework in Figure 1.11 and assign velocity $v_2 \in \mathbb{R}^2$ to vertex 2 such that it is any vector *perpendicular* to $p_1 - p_2$ and $p_2 - p_3$ while keeping all other velocities at zero, i.e., $v = [0, v_2, 0, 0, 0, 0]$. It is easy to see that $R(p)v = 0$ in this case although v is not due to a rigid body motion. Therefore, the framework is infinitesimally flexible.[6] This trial-and-error method becomes cumbersome as the complexity of the framework increases or when the framework is generic. More important, it cannot be easily implemented computationally.

Fortunately, there is an easy way of checking whether a framework is infinitesimally rigid via the rank of the rigidity matrix. Before presenting this useful result, we need the following facts. First, the rank of an $m \times n$ matrix is equal to n minus the dimension of its null space. Second, the number of independent rigid body motions for a generic framework in $\mathbb{R}^m$ is $\frac{m(m+1)}{2}$ [18]. This means that there exist *three* independent rigid body motions in $\mathbb{R}^2$ (translation along x, translation along y, and rotation about z) and *six* in $\mathbb{R}^3$ (translations along x, y, z and rotations about x, y, z).

Given the above, we deduce that $\dim(\text{Null}(R(p))) \geq \frac{m(m+1)}{2}$ and therefore $\text{rank}(R(p)) \leq nm - \frac{m(m+1)}{2}$ [15]. From Definition 1.7, we know that for infinitesimally rigid frameworks $\dim(\text{Null}(R(p))) = \frac{m(m+1)}{2}$, which gives us the following result.

Result 1.1 *[14]* A (generic) framework in $\mathbb{R}^m$ is infinitesimally rigid if and only if

$$\text{rank}(R(p)) = nm - \frac{m(m+1)}{2}. \tag{1.21}$$

Therefore, $\text{rank}(R(p)) = 2n - 3$ in $\mathbb{R}^2$ and $\text{rank}(R(p)) = 3n - 6$ in $\mathbb{R}^3$.

Example 1.6 Consider the framework in Figure 1.6b whose rigidity matrix is given in (1.17), and let $p_1 = [1, 1]$, $p_2 = [0, 1]$, $p_3 = [0, 0]$, $p_4 = [1, 0]$, and

---

6 This framework is, however, rigid since it is nongeneric.

$p_5 = [2, 0]$. Then,

$$R(p) = \begin{bmatrix} 1 & 0 & -1 & 0 & 0 & 0 & 0 & 0 & 0 & 0 \\ 0 & 0 & 0 & 1 & 0 & -1 & 0 & 0 & 0 & 0 \\ 0 & 0 & 0 & 0 & -1 & 0 & 1 & 0 & 0 & 0 \\ 0 & 1 & 0 & 0 & 0 & 0 & 0 & -1 & 0 & 0 \\ 0 & 0 & -1 & 1 & 0 & 0 & 1 & -1 & 0 & 0 \\ 0 & 0 & 0 & 0 & 0 & 0 & -1 & 0 & 1 & 0 \end{bmatrix}$$ (1.22)

and $\text{rank}(R(p)) = 6$, which is less than $2n - 3 = 7$. Thus, the framework is infinitesimally flexible. If we now remove vertex 5 and edge 6 (connecting vertices 4 and 5) from Figure 1.6b, we obtain a framework similar to the one in Figure 1.9b. Its rigidity matrix will be the 5×8 submatrix of (1.22) that results from deleting the last row and the last two columns. In this case, $\text{rank}(R(p)) = 5$ $(= 2n - 3 = 2 \times 4 - 3)$ and the framework is infinitesimally rigid in $\mathbb{R}^2$.

Example 1.7 Consider the three-dimensional framework in Figure 1.12 where $E = \{(1, 2), (2, 3), (1, 3), (1, 4), (2, 4), (3, 4)\}$ and $p_1 = [0, 0, 0]$, $p_2 = [0, 1, 0]$, $p_3 = [1, 0, 0]$, and $p_4 = [0.5, 0.5, 1]$. The rigidity matrix is given by

$$R(p) = \begin{bmatrix} 0 & -1 & 0 & 0 & 1 & 0 & 0 & 0 & 0 & 0 & 0 & 0 \\ 0 & 0 & 0 & -1 & 1 & 0 & 0 & 1 & -1 & 0 & 0 & 0 \\ -1 & 0 & 0 & 0 & 0 & 0 & 1 & 0 & 0 & 0 & 0 & 0 \\ -0.5 & -0.5 & -1 & 0 & 0 & 0 & 0 & 0 & 0 & 0.5 & 0.5 & 1 \\ 0 & 0 & 0 & -0.5 & 0.5 & -1 & 0 & 0 & 0 & 0.5 & -0.5 & 1 \\ 0 & 0 & 0 & 0 & 0 & 0 & 0.5 & -0.5 & -1 & -0.5 & 0.5 & 1 \end{bmatrix}$$

whose rank is 6 $(= 3n - 6)$ so the framework is infinitesimally rigid in $\mathbb{R}^3$.

As discussed earlier, the rigidity property of a generic framework $F = (G, p)$ is invariant under small perturbations on p. With the intent of formalizing this idea, we introduce the following result.

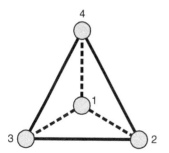

Figure 1.12 A tetrahedron framework.

Result 1.2 Consider two frameworks $F = (G, p)$ and $\overline{F} = (G, \overline{p})$ sharing the same graph. If F is infinitesimally rigid and dist$(\overline{p}, \text{Iso}(F)) \leq \varepsilon$ where ε is a sufficiently small positive constant, then $\overline{F}$ is also infinitesimally rigid.

Proof: Let $\hat{F} = (G, \hat{p}) \in \text{Iso}(F)$ be such that

$$\text{dist}(\overline{p}, \text{Iso}(F)) = \inf_{x \in \text{Iso}(F)} \|\overline{p} - x\| = \|\overline{p} - \hat{p}\|. \tag{1.23}$$

Obviously, $\hat{F}$ is infinitesimally rigid since it is isomorphic to F. Therefore, rank$(R(\hat{p})) = 2n - 3$ according to Result 1.1, and there exists a $(2n - 3) \times (2n - 3)$ submatrix of $R(\hat{p})$, $R_s(\hat{p})$, such that $\det[R_s(\hat{p})] \neq 0$. The submatrix $R_s(\hat{p})$ has nonzero elements of the form $(\hat{p}_i - \hat{p}_j)^T$, $(i, j) \in E$. Since dist$(\overline{p}, \text{Iso}(F)) = \|\overline{p} - \hat{p}\| \leq \varepsilon$, it is not difficult to show that $[\overline{p}_i]_k = [\hat{p}_i]_k + \gamma_{ik}$ where γ_{ik} is a sufficiently small positive constant. Thus, the nonzero elements of $R_s(\overline{p})$ have the form $[\overline{p}_i]_k - [\overline{p}_j]_k = [\hat{p}_i]_k - [\hat{p}_j]_k + \gamma_{ik} - \gamma_{jk}$, which are continuously dependent on $\hat{p}$. Since the eigenvalues of a matrix depend continuously on its elements [19], and the determinant of a matrix is the product of its eigenvalues, it follows that the determinant continuously depends on the elements of the matrix. Thus, for sufficiently small γ_{ik}, we have that $\det[R_s(\overline{p})] \neq 0$ and rank$(R_s(\overline{p})) = \text{rank}(R_s(\hat{p})) = 2n - 3$. Now, since $R_s(\overline{p})$ is a full rank submatrix of $R(\overline{p})$, we know rank$(R(\overline{p})) = 2n - 3$, so $\overline{F}$ is infinitesimally rigid.

1.3.5 Minimal Rigidity

It is obvious that adding edges to a graph does not destroy rigidity. A natural question is then: What is the *minimum* number of edges that ensures rigidity? This is important in practice because it ensures that a formation of multiple agents is rigid with the minimum number of sensing and communication links.[7]

Definition 1.8 *[6]* A graph is minimally rigid if it is rigid and the removal of a single edge causes it to lose rigidity.

Example 1.8 The graph in Figure 1.13a is minimally rigid because the removal of any single edge will make it flexible. On the other hand, any single edge can be removed from the graph in Figure 1.13b and it remains rigid.

Just like with infinitesimal rigidity, we want a checkable condition for minimal rigidity. Fortunately, the following condition exists.

---

7 The disadvantage of minimal rigidity is that it lacks robustness to the loss of a sensor or communication link since redundant edges are removed from the graph [6].

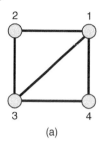 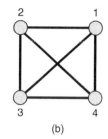

Figure 1.13 (a) Minimally and (b) nonminimally rigid graphs.

(a)   (b)

Result 1.3 *[6]* A rigid graph is minimally rigid if and only if $l = mn - \frac{m(m+1)}{2}$.

It is interesting to note the recurrence of the term $mn - \frac{m(m+1)}{2}$ in Results 1.1 and 1.3. This leads to the following corollary, which will be very useful for the control designs in this book.

Corollary 1.1 If a framework is infinitesimally and minimally rigid, then its rigidity matrix has *full row rank* and $R(p)R^{\mathsf{T}}(p)$ is positive definite.

Proof: Given that the rigidity matrix has l rows and rank$(R(p)) = nm - \frac{m(m+1)}{2}$ and $l = mn - \frac{m(m+1)}{2}$ for infinitesimally and minimally rigid frameworks, we know $R(p)$ is full row rank. Now, let $y = R^{\mathsf{T}}(p)x$ and

$$V = y^{\mathsf{T}}y \geq 0.$$

Note that V is positive definite with respect to y and positive semi-definite with respect to x. Given that rank$(R(p)) = $ rank$(R(p)R^{\mathsf{T}}(p))$ and $R(p)$ is full row rank, we know $R(p)R^{\mathsf{T}}(p)$ is invertible and has no zero eigenvalues. Therefore, V is positive definite with respect to x and $R(p)R^{\mathsf{T}}(p)$ is a positive definite matrix.

1.3.6 Framework Ambiguities

Consider that a graph $G = (V, E)$ and the length of each edge (i.e., $\|p_i - p_j\|$, $(i, j) \in E$) are given. We want to know all frameworks $F = (G, p)$ that are consistent with this data *excluding* isometries. Can a framework have multiple (non-isomorphic) realizations? There are several sources of nonunique realizations. The trivial one is flexibility, such as in Figure 1.6. Since edge (4,5) can continuously rotate about vertex 4, an infinite number of equivalent frameworks exist that satisfy the edge constraints. Infinitesimally rigid frameworks can also suffer from nonuniqueness. Two types of nonuniqueness can occur in this case, but unlike flexibility they lead to a finite number of equivalent frameworks [20, 21]:

- Flip ambiguity: This occurs when a graph in $\mathbb{R}^m$ has a set of vertices lying in a $(m-1)$-dimensional subspace about which a portion of the graph can be reflected. This is illustrated in Figure 1.14a, where edges $(1, 2)$ and $(1, 4)$ are reflected over the "mirror" edge $(2, 4)$. Notice that there cannot be any edges between the portions of the graph separated by the mirror edge. Multiple flips can happen in a given framework. For example, if edges $(2, 3)$ and $(3, 4)$ also reflect over edge $(2, 4)$, one ends up with the completely reversed framework.
- Flex ambiguity: This occurs in minimally rigid graphs since the removal of an edge allows a portion of the graph to flex. If the removed edge is reinserted after a flexing, one may obtain an equivalent framework with a different configuration. An example is shown in Figure 1.14b, where the temporary remove of edge $(2, 3)$ allows edges $(2, 5)$, $(1, 2)$, and $(1, 4)$ to flex.

This leads us to the following definition.

Definition 1.9   If two infinitesimally rigid frameworks are equivalent but not congruent, then they are said to be *ambiguous*.

We denote the set of all ambiguities of an infinitesimally rigid framework F by Amb(F). We assume that all frameworks in Amb(F) are also infinitesimally rigid.[8] The existence of framework ambiguities is problematic for formation control since the control law cannot distinguish the actual framework from an ambiguous one if only the edge lengths (i.e., inter-agent distances) are being controlled. In such a case, one solution is to initialize the agents sufficiently close to the desired framework to avoid their convergence to an ambiguous

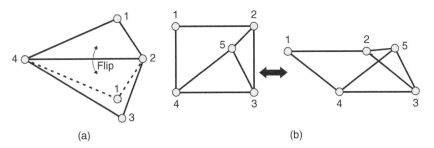

(a)   (b)

Figure 1.14 Examples of nonunique realizations for infinitesimally rigid frameworks: (a) flip ambiguity and (b) flex ambiguity.

---

8 This assumption is reasonable and, in fact, holds almost everywhere; see [6] and Theorem 3 of [22] for details.

framework. From a control theory standpoint, this means that stability results will be *local* in nature rather than global. The following corollary to Result 1.2 will be helpful in establishing the stability sets in this book.

Corollary 1.2   Let $F = (G, p)$ and $\overline{F} = (G, \overline{p})$ be two frameworks sharing the same graph, and consider the function

$$\Psi(\overline{F}, F) = \sum_{(i,j) \in E} (\|\overline{p}_i - \overline{p}_j\| - \|p_i - p_j\|)^2. \tag{1.24}$$

If F is infinitesimally rigid and $\Psi(\overline{F}, F) \leq \delta$ where δ is a sufficiently small positive constant, then $\overline{F}$ is also infinitesimally rigid.

Proof: First, note that $\Psi(\overline{F}, F) = 0$ implies that $\overline{F} \in \text{Iso}(F)$ or $\overline{F} \in \text{Amb}(F)$. Therefore, $\Psi(\overline{F}, F) \leq \delta$ implies that there is a sufficiently small positive constant ε such that $\text{dist}(\overline{p}, \text{Iso}(F)) \leq \varepsilon$ or $\text{dist}(\overline{p}, \text{Amb}(F)) \leq \varepsilon$. From Result 1.2, we know that $\overline{F}$ is infinitesimally rigid if $\text{dist}(\overline{p}, \text{Iso}(F)) \leq \varepsilon$. Since the elements of $\text{Amb}(F)$ are infinitesimally rigid, the proof of Result 1.2 can be followed with $\text{Iso}(F)$ replaced by $\text{Amb}(F)$ to show that $\text{dist}(\overline{p}, \text{Amb}(F)) \leq \varepsilon$ implies $\overline{F}$ is infinitesimally rigid.

1.3.7   Global Rigidity

A final variation of graph rigidity is the concept of *global rigidity*.

Definition 1.10   *[13, 23]*   A framework (G, p) is globally rigid if every framework that is equivalent to (G, p) is congruent to (G, p). In more technical terms, the framework is globally rigid if $\phi^{-1}(\phi(p)) = \phi_K^{-1}(\phi_K(p))$ where ϕ_K denotes the edge function of the framework (K, p) and K is the complete graph with the same number of vertices as G.

Global rigidity is a stricter concept than (plain) rigidity since a framework can be rigid but not globally rigid. In such cases, the framework can be converted to a globally rigid one by the addition of edges (see Figure 1.15 for an example).

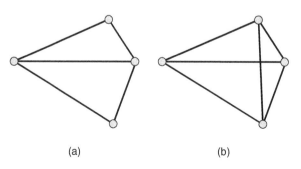

Figure 1.15 (a) Rigidity versus (b) global rigidity.

(a)   (b)

Global rigidity avoids the occurrence of framework ambiguities since it ensures that the shape of the framework is unique.

1.4 Formation Control Problems

This book will address a number of different formation control problems for multi-agent systems. In this section, we formally introduce these problems.

Consider a system of n mobile agents where $q_i \in \mathbb{R}^m$ is the position of the ith agent relative to an Earth-fixed coordinate frame, and $u_i \in \mathbb{R}^m$ is the corresponding control input. In subsequent chapters, u_i will be a velocity-, acceleration-, or actuator-level input depending on the mathematical model used to describe the agent motion.

Let the *desired* formation for the agents be represented by an *infinitesimally and minimally rigid* framework $F^* = (G^*, q^*)$ where $G^* = (V^*, E^*)$ is the formation graph, $\dim(V^*) = n$, $\dim(E^*) = l$, and $q^* = [q_1^*, \ldots, q_n^*]$. The constant desired distance between agents i and j is given by[9]

$$d_{ij} = \|q_i^* - q_j^*\| > 0, \quad i, j \in V^*. \tag{1.25}$$

In practice, the geometric shape/structure of the desired formation is dictated by the mission to be accomplished by the agents. When translating the desired shape into a framework, one needs to include enough edges to ensure that F^* is indeed infinitesimally and minimally rigid.

The *actual* formation of the agents is represented by the framework $F(t) = (G_s, q(t))$ where G_s represents the sensor graph and $q = [q_1, \ldots, q_n]$. It is important to clarify the difference between the formation graph G^* and the sensor graph G_s, which in general need not be the same. G^* indicates the minimum number of inter-agent distances that need to be controlled for the desired formation to be successfully reached. On the other hand, G_s indicates the agent pairs that can sense and/or communicate with each other.

We make the following assumptions regarding the desired and actual formations:

Assumption 1 The set where the agents achieve the desired formation is nonempty, i.e., there exist q^* such that $\phi(q^*) = d$ where $d = [\ldots, d_{ij}^2, \ldots] \in \mathbb{R}^l$ and ϕ was defined in (1.7).

Assumption 2 The formation and sensor graphs are the same, i.e., $G_s = G^*$. Furthermore, inter-agent connectivity is always maintained in the sense that agent i is always within the sensing/communication range of agent j, $\forall j \in \mathcal{N}_i(E^*)$. In other words, G^* is fixed.[10]

---

9 In Sections 2.5 and 3.5 we will consider the case where the desired distances are time varying.
10 Connectivity maintenance prevents the occurrence of flex ambiguities since temporary loss of edges cannot happen.

Assumption 3   At $t = 0$, the agents do not satisfy the desired inter-agent distance constraints, i.e., $\|q_i(0) - q_j(0)\| \neq d_{ij}$, $i, j \in V^*$.

Assumption 4   The only position information being measured is the *relative* position of agent pairs in E^*, $q_i - q_j$, $(i, j) \in E^*$.[11] That is, the global position of the agents, q_i, $i = 1, \dots, n$, are not available to the control.

We will deal with three types of control problems: formation acquisition, formation maneuvering, and target interception.

Problem 1   *(Formation Acquisition)*   The goal is for the agents to acquire and maintain a pre-defined geometric shape in space. The control objective for formation acquisition, which serves as the *common*, primary objective for the other two problems, can be mathematically described as to design u_i such that

$$F(t) \to \text{Iso}(F^*) \text{ as } t \to \infty. \tag{1.26}$$

Note that (1.26) is equivalent to

$$\|q_i(t) - q_j(t)\| \to d_{ij} \text{ as } t \to \infty, \quad i, j \in V^*. \tag{1.27}$$

Since only the inter-agent distances are to be directly controlled, the actual formation can converge to any isometry of F^*. That is, the meaning of (1.26) is that the formation will converge to one framework in the set $\text{Iso}(F^*)$ with the specific one being determined by the initial position of the agents, $q_i(0)$, $i = 1, \dots, n$.   △

Problem 2   *(Formation Maneuvering)*   The agents are required to simultaneously acquire a formation (i.e., satisfy (1.26)) and maneuver cohesively according to some pre-defined trajectory. Thus, the secondary objective is

$$\dot{q}_i(t) - v_{di}(t) \to 0 \text{ as } t \to \infty, \quad i = 1, \dots n \tag{1.28}$$

where $v_{di} \in \mathbb{R}^3$ represents the desired rigid body velocity for the swarm of agents. That is, the fixed-shape, desired formation evolves in space as a virtual rigid body undergoing translation and/or rotation.

In practice, the selection of v_{di} is mission dependent. For example, it could be related to a path planning algorithm that provides an optimal solution to the coverage problem where agents cooperatively maximize the coverage area of a given mission under certain time and/or fuel consumption constraints.

When v_{di} only includes a translation velocity, the formation maneuvering problem is also called *flocking*. For the case where v_{di} has a rotational component, we assign the nth agent (without lost of generality) to be the "leader" while

---

11 As we will see in the following chapters, the control could also be a function of other, nonposition-related variables depending on the agent model and formation problem being solved.

Figure 1.16 Example of the construction of F^*: a tetrahedron formation where $\mathcal{L}$ stands for leader and $\mathcal{F}$ for follower.

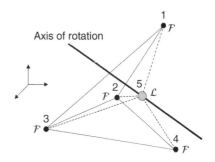

the remaining agents are "followers". This assignment is for the sole purpose of one agent serving as a reference point for the axis of rotation of the virtual rigid body (i.e., point O in Figure 1.7). Therefore, F^* should be constructed with the following additional conditions:

- $q_n^* \in \text{conv}\{q_1^*, ..., q_{n-1}^*\}$;
- $(i, n) \in E^*$, $i = 1, \ldots, n-1$, i.e., there is an edge between each follower and the leader.

An example of F^* is illustrated by the 3D formation in Figure 1.16 where the leader is located in the interior of the tetrahedron. The axis of rotation passes through the leader, which is inside the tetrahedron. Since $n = 5$, we need $3n - 6 = 9$ for the framework to be minimally rigid. The solid lines indicate edges that form the faces of the tetrahedron while the dashed lines are edges in its interior. Notice that edge $(1, 4)$ is not necessary.

The association of a leader agent (instead of a *virtual* leader) with the axis of rotation is done for convenience (not necessity) since the leader's relative position to the followers can be measured and it will not have to undergo any rotation. Note that if one uses a virtual leader, its location would have to be known in order to calculate its position relative to the agents (see (1.10)). This in turn would require extra measurements and/or calculations. △

Problem 3 *(Target Interception)* The agents should intercept and surround a (possibly evading) moving target with a pre-defined formation. Here, we will also use the leader–follower approach by taking the nth agent to be the leader while the remaining agents are followers. The control protocol will consist of: (i) selecting F^* such that $q_n^* \in \text{conv}\{q_1^*, ..., q_{n-1}^*\}$[12], (ii) the leader chasing the target, and (iii) the followers tracking the leader while maintaining the desired formation. Thus, if $q_T \in \mathbb{R}^m$ denotes the target position, the secondary

---

12 Unlike formation maneuvering with rotation, we do not need $(i, n) \in E^*$, $i = 1, \ldots, n-1$ for target interception.

objective for this problem is that $q_T(t)$ approach $\text{conv}\{q_1(t), q_2(t), ..., q_{n-1}(t)\}$ as time evolves, which (with abuse of notation) we express as

$$q_T(t) \in \text{conv}\{q_1(t), q_2(t), ..., q_{n-1}(t)\} \text{ as } t \to \infty. \tag{1.29}$$

$\triangle$

1.5   Book Overview and Organization

The subsequent chapters of this book will introduce a class of graph rigidity-based formation controllers for multi-agent systems. The control laws primarily stabilize the inter-agent distance dynamics to desired distances (formation acquisition) using system models of increasing complexity and accuracy: single integrator, double integrator, and robotic vehicle kinematics and dynamics. For problems of formation maneuvering and target interception, the control will contain additional terms to account for these secondary objectives. The inter-agent distance dynamics and, consequently, the control laws will be formulated in terms of the rigidity matrix. The stability of these dynamics in closed loop with the control will depend on the rigidity matrix having full row rank, thus the need for requiring that the desired formation F^* is infinitesimally and minimally rigid (see Corollary 1.1). The book includes theoretical, simulation, and experimental results, and is organized as follows.

We naturally begin the presentation with the simple single-integrator model in Chapter 2. Here, we first design a formation acquisition controller that ensures that the desired formation is exponentially stable. The control is distributed in the sense that the control input of each agent is only a function of the relative position of neighboring agents in the graph. The formation acquisition controller is the *basic* control term since it will appear in all other control algorithms developed in the book. Building upon this result, we show how the formation acquisition control can be augmented with a term to enable the agents to perform formation maneuvering or target interception simultaneously with formation acquisition. In the formation maneuvering3 problem, the swarm (group) velocity is first assumed to be known to all agents. We then consider a variant of this problem where the agents are just flocking with a constant velocity; however, this velocity is known only to a subset of agents. An observer is introduced to estimate the flocking velocity. Next, the idea behind the control design for formation maneuvering is extended to the target interception problem. We assume the target's relative position to the leader agent is known and broadcast to the followers; however, the target's velocity is *unknown* to all agents. To deal with this uncertainty, the target interception component of the control law will contain a continuous dynamic robust mechanism, inspired by the work in [24], to estimate the target velocity. Our stability analysis for both problems provides exponential formation acquisition and asymptotic formation maneuvering or target interception. We close this chapter with the case where the desired formation is *dynamic* rather than static. This is motivated by situations, such as obstacle avoidance and limited communication range/bandwidth, where the formation size and/or geometric shape need to vary in time. We also briefly discuss how the control law can be modified to perform formation maneuvering on top of tracking the time-varying formation.

In Chapter 3, the results from Chapter 2 are extended to the double-integrator model. The backstepping control technique is a natural tool for solving this problem since it allows us to treat the velocity-level inputs designed in Chapter 2 as *fictitious* control inputs, which are to be tracked by the new, acceleration-level inputs. For formation acquisition, the control input of each agent is dependent on the relative position and relative velocity of neighboring agents in the graph and the agents' own velocity. The formation maneuvering control is dependent on the desired swarm acceleration in addition to the velocity. In the target interception problem, we assume the target's relative position to the leader and velocity are known and can be broadcast to the followers; however, the target's *acceleration* is unknown to all agents. To deal with this uncertainty, the target interception component of the control law will contain a variable structure-type term to compensate for the unknown target acceleration. The stability results of Chapter 2 are preserved in these results.

In Chapter 4, we account for the nonlinear kinematics and dynamics of the agents during the control design process. The control here is limited to the formation acquisition problem. We consider a class of robotic vehicles moving on the plane, such as unicycle robots, marine (surface) vessels, underwater vehicles with constant depth, and aircraft with constant altitude. We first deal with the nonholonomic kinematic equation only, and show how the formation acquisition control from Chapter 2 can be indirectly used. We then account for the vehicle dynamics by transforming the equations of motion into an Euler–Lagrange-like system so we can exploit its structural properties in the control design and stability analysis. The backstepping technique is again applied to incorporate the velocity-level inputs from Chapter 2 into the torque-level control law in a rigorous manner. We consider the cases where the dynamics are fully known as well as subject to parametric uncertainty. In the latter case, we show how the control law can be redesigned with parameter adaptation to compensate for the uncertainties.

The book culminates with the demonstration of the experimental implementation of the controllers from Chapters 2, 3, and 4 on an actual car-like robotic platform. Three customized, unmanned ground vehicles (UGVs) were used for this purpose. We show how the high-level, formation algorithms can be embedded in the motor-level commands to the UGVs in order to acquire and maneuver a given triangular formation.

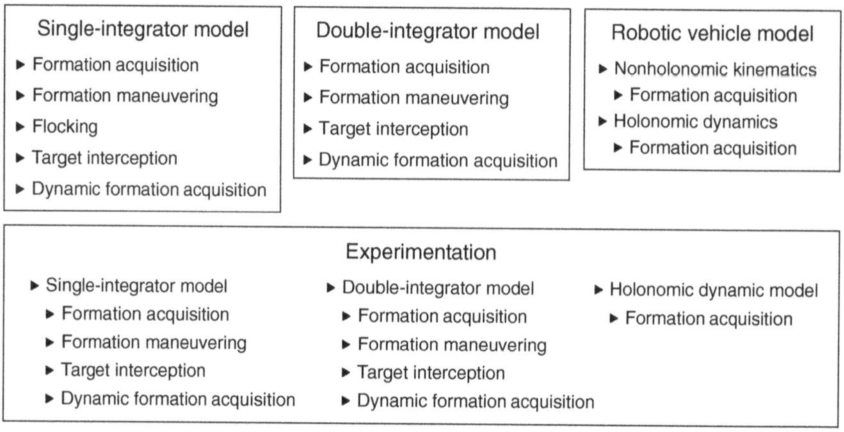

Single-integrator model	Double-integrator model	Robotic vehicle model
► Formation acquisition	► Formation acquisition	► Nonholonomic kinematics
► Formation maneuvering	► Formation maneuvering	► Formation acquisition
► Flocking	► Target interception	► Holonomic dynamics
► Target interception	► Dynamic formation acquisition	► Formation acquisition
► Dynamic formation acquisition		

Experimentation

► Single-integrator model	► Double-integrator model	► Holonomic dynamic model
► Formation acquisition	► Formation acquisition	► Formation acquisition
► Formation maneuvering	► Formation maneuvering	
► Target interception	► Target interception	
► Dynamic formation acquisition	► Dynamic formation acquisition	

Figure 1.17 Overview of the book organization.

An overview of the book organization is shown in Figure 1.17.

1.6 Notes and References

In-depth coverage of rigid graph theory can be found in, for example, [13–18, 25–29]. Rigid graphs were first applied to formation control in [30]. Some early work on the application of graph theory to multi-agent formation appeared in [31–35]. An overview of rigid graph theory and its application to sensing, communication, and control architectures for formations of autonomous vehicles was presented in [6]. A number of books have been published in the past few years that deal with the general topic of cooperative control of multi-agent systems [2–4, 37–39]. A recent survey of multi-agent formation control was provided in [40].

2

Single-Integrator Model

This chapter will set the foundation for the formation control designs presented in the book. We use here a very simple model for the motion of the agents known as the *single-integrator model*, which only includes two variables: position and velocity. This is a simplified kinematic model for omnidirectional robots (e.g., mobile robots with Swedish wheels [41]). Specifically, we consider a system of n agents governed by the first-order differential equation

$$\dot{q}_i = u_i, \quad i = 1, \dots, n \tag{2.1}$$

where $q_i \in \mathbb{R}^m$ is the position and $u_i \in \mathbb{R}^m$ is the velocity-level control input of the ith agent with respect to an Earth-fixed coordinate frame. The name "single integrator" originates from the fact that the transfer function matrix of (2.1) is

$$G_i(s) = \frac{1}{s} I_m \tag{2.2}$$

where s is the Laplace variable, i.e., the inputs and outputs are separated by one integrator.

Formation controllers based on (2.1) are called high-level control laws because they are often embedded in controllers designed for more refined agent models. Therefore, the control laws introduced in this chapter will form the basis for all subsequent designs.

2.1 Formation Acquisition

We begin with the formation acquisition problem defined in Section 1.4. Given (2.1), we seek to design $u_i = u_i(q_i - q_j, d_{ij})$, $i = 1, \dots, n$ and $j \in \mathcal{N}_i(E^*)$, where $\mathcal{N}_i(\cdot)$ was defined in (1.2) to achieve the control objective described by (1.26) (or equivalently (1.27)).

It is appropriate at this point to elaborate on an issue mentioned at the end of Section 1.3.6 regarding framework ambiguities. The inputs u_i, $i = 1, \dots, n$

Formation Control of Multi-Agent Systems: A Graph Rigidity Approach, First Edition.
Marcio de Queiroz, Xiaoyu Cai, and Matthew Feemster.
© 2019 John Wiley & Sons Ltd. Published 2019 by John Wiley & Sons Ltd.
Companion website: www.wiley.com/go/dequeiroz/formation_control

will *directly* control the distances $\|q_i - q_j\|$, $(i,j) \in E^*$. Therefore, they can only directly ensure that

$$\|q_i(t) - q_j(t)\| \to d_{ij} \text{ as } t \to \infty, \quad (i,j) \in E^*, \tag{2.3}$$

which is equivalent to

$$\phi(q(t)) \to \phi(q^*) = d \text{ as } t \to \infty. \tag{2.4}$$

Note that (2.3) is different than (1.27) since it is only defined for $(i,j) \in E^*$ while (1.27) is defined for all $i,j \in V^*$. This is potentially problematic since (with abuse of notation) $\phi(\text{Iso}(F^*)) = \phi(\text{Amb}(F^*))$. Therefore, the control scheme will need to avoid the possibility that $F(t) \to \text{Amb}(F^*)$ as $t \to \infty$. This will be accomplished by initializing the agents sufficiently close to $\text{Iso}(F^*)$ in the sense that $\text{dist}(q(0), \text{Iso}(F^*)) < \text{dist}(q(0), \text{Amb}(F^*))$.

To simplify the notation in the following derivations, we define the relative position of two agents as

$$\tilde{q}_{ij} = q_i - q_j \tag{2.5}$$

and let $\tilde{q} = [\dots, \tilde{q}_{ij}, \dots] \in \mathbb{R}^{ml}$, $(i,j) \in E^*$ with the same ordering of terms as the edge function (1.7). The distance error is given by

$$e_{ij} = \|\tilde{q}_{ij}\| - d_{ij}. \tag{2.6}$$

Note that (1.27) is equivalent to $e_{ij}(t) \to 0$ as $t \to \infty, i,j \in V^*$. The distance error dynamics can be derived from (2.6) and (2.1) as

$$\begin{aligned}
\dot{e}_{ij} &= \frac{d}{dt}\left(\sqrt{\tilde{q}_{ij}^\top \tilde{q}_{ij}}\right) \\
&= (\tilde{q}_{ij}^\top \tilde{q}_{ij})^{-\frac{1}{2}} \tilde{q}_{ij}^\top (u_i - u_j) \\
&= \frac{\tilde{q}_{ij}^\top (u_i - u_j)}{e_{ij} + d_{ij}}.
\end{aligned} \tag{2.7}$$

Let

$$z_{ij} = \|\tilde{q}_{ij}\|^2 - d_{ij}^2, \tag{2.8}$$

which can be rewritten as

$$z_{ij} = e_{ij}(e_{ij} + 2d_{ij}) \tag{2.9}$$

using (2.6). Given that $\|\tilde{q}_{ij}\| \geq 0$ (or equivalently, $e_{ij} \geq -d_{ij}$), it is not difficult to see that $z_{ij} = 0$ if and only if $e_{ij} = 0$. We now introduce the following Lyapunov function candidate

$$W(e) = \frac{1}{4} \sum_{(i,j) \in E^*} z_{ij}^2 = \frac{1}{4} z^\top z \tag{2.10}$$

where $e = [\dots, e_{ij}, \dots] \in \mathbb{R}^l$ and $z = [\dots, z_{ij}, \dots] \in \mathbb{R}^l$, $(i,j) \in E^*$ are ordered as (1.7). This function is positive definite in e and its level surfaces, $W(e) = c$ for some $c > 0$, are closed since $e_{ij} \geq -d_{ij}$.

The time derivative of (2.10) along (2.7) is given by

$$\dot{W} = \sum_{(i,j) \in E^*} e_{ij}(e_{ij} + 2d_{ij})\tilde{q}_{ij}^\top(u_i - u_j). \tag{2.11}$$

Using (1.15), (1.16), and (2.9), (2.11) can be conveniently written as[1]

$$\dot{W} = z^\top R(\tilde{q})u \tag{2.12}$$

where $u = [u_1, \dots, u_n] \in \mathbb{R}^{mn}$ is the stacked vector of control inputs.

Before presenting the main result, we introduce a lemma that establishes the relationship between Corollary 1.2 and the level surfaces of the Lyapunov function candidate.

Lemma 2.1 For nonnegative constants c and δ, the level set $W(e) \leq c$ is equivalent to $\Psi(F, F^*) \leq \delta$ where Ψ and W were defined in (1.24) and (2.10), respectively.

Proof: First, from (1.24), (1.25), (2.5), and (2.6), we have that

$$\Psi(F, F^*) = \sum_{(i,j) \in E^*} (\|q_i - q_j\| - \|q_i^* - q_j^*\|)^2$$

$$= \sum_{(i,j) \in E^*} (\|q_i - q_j\| - d_{ij})^2$$

$$= \sum_{(i,j) \in E^*} e_{ij}^2. \tag{2.13}$$

From (2.10), we know $W(e) \leq c$ implies that e_{ij}, $(i,j) \in E^*$ is bounded. This boundedness along with (2.13) implies $\Psi(F, F^*) \leq \delta$ where δ is some nonnegative constant. Now, given $\Psi(F, F^*) \leq \delta$, it follows from (2.13) that e_{ij} is bounded for $(i,j) \in E^*$. This implies z_{ij}, $(i,j) \in E^*$ is bounded, and $W(e) \leq c$ where c is some nonnegative constant. $\square$

The control law for solving the formation acquisition problem is given in the following theorem. Its structure is based on (2.12) and Lyapunov stability theory. Specifically, the goal is to make the time derivative of the Lyapunov function candidate *negative definite* [8].

---

1 Although the argument of the rigidity matrix function is commonly written as q, it is obvious from (1.7) and (1.15) that R is dependent on $\tilde{q}$ only. Henceforth, we write $R(\tilde{q})$ so it is clear that the matrix is a function of the relative position.

Theorem 2.1   Consider the formation $F(t) = (G^*, q(t))$, and let the initial conditions of the error dynamics be such that $e(0) \in \Omega_1 \cap \Omega_2$ where

$$\Omega_1 = \{e \in \mathbb{R}^l \mid \Psi(F, F^*) \leq \delta\},$$

$$\Omega_2 = \{e \in \mathbb{R}^l \mid \text{dist}(q, \text{Iso}(F^*)) < \text{dist}(q, \text{Amb}(F^*))\}, \tag{2.14}$$

and δ is a sufficiently small positive constant. The control law[2]

$$u = u_a := -k_v R^\top(\tilde{q})z, \tag{2.15}$$

where $k_v > 0$ is a user-defined control gain, renders $e = 0$ exponentially stable and ensures (1.26) is satisfied.

Proof: Given that F^* and $F(t)$ have the same number of edges and that F^* is minimally rigid by design, then $F(t)$ is minimally rigid for all $t \geq 0$. Substituting (2.15) into (2.12) yields

$$\dot{W} = -k_v z^\top R(\tilde{q}) R^\top(\tilde{q}) z. \tag{2.16}$$

Since F^* is infinitesimally rigid, we know from Corollary 1.2 that $F(t)$ is infinitesimally rigid for $e(t) \in \Omega_1$. Therefore, we know $F(t)$ is infinitesimally and minimally rigid for $e(t) \in \Omega_1$, so we can invoke Corollary 1.1 to state

$$\dot{W} \leq -k\lambda_{\min}(RR^\top)z^\top z = -4k\lambda_{\min}(RR^\top)W \quad \text{for} \quad e(t) \in \Omega_1 \tag{2.17}$$

where (2.10) was used. From (2.17), we know that $\dot{W}(t) \leq 0$ for all $t \geq 0$; hence, $W(t)$ is nonincreasing for all $t \geq 0$. Then, since $e(t) \in \Omega_1$ is equivalent to $e(t) \in \{e \in \mathbb{R}^{3n} \mid W(e) \leq c\}$ from Lemma 2.1, a sufficient condition for (2.17) is given by

$$\dot{W} \leq -4k\lambda_{\min}(RR^\top)W \quad \text{for} \quad e(0) \in \Omega_1. \tag{2.18}$$

From the form of (2.18) and the fact that W is positive definite in e, we can invoke Corollary C.1 to conclude that $e = 0$ is exponentially stable [8] for $e(0) \in \Omega_1$. Given that e is only defined for $(i, j) \in E^*$, the exponential stability of $e = 0$ implies that $F(t) \to \text{Iso}(F^*)$ or $F(t) \to \text{Amb}(F^*)$ as $t \to \infty$. If we choose $e(0) \in \Omega_1 \cap \Omega_2$, we have from (2.14) that

$$\text{dist}(q(0), \text{Iso}(F^*(0))) < \text{dist}(q(0), \text{Amb}(F^*(0))). \tag{2.19}$$

Due to (2.19), the energy-like function $W(t)$ would need to increase for a period of time for $F(t) \to \text{Amb}(F^*)$ as $t \to \infty$, which is a contradiction since (2.18) establishes that $W(t)$ is nonincreasing for all $t \geq 0$. Therefore, we know $F(t) \to \text{Iso}(F^*)$ as $t \to \infty$ for $e(0) \in \Omega_1 \cap \Omega_2$. This argument is conceptually illustrated by Figure 2.1, where the ball, representing $F(t)$, would have to overcome the energy barrier to reach $\text{Amb}(F^*)$.   □

---

2 The variable u_a in (2.15) denotes the basic formation acquisition control term that will be embedded in all control algorithms developed in the book.

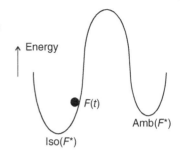

Figure 2.1 Energy landscape showing the two equilibrium points, Iso(F^*) and Amb(F^*), at the bottom of each well.

The initial condition $e(0) \in \Omega_1 \cap \Omega_2$ in Theorem 2.1 is a sufficient condition for the actual formation $F(t)$ to (i) remain infinitesimally rigid for all time and (ii) be closer to a framework in Iso(F^*) at $t = 0$ than to one in Amb(F^*) in order to avoid converging to an ambiguous framework. The former constraint is satisfied by $e(0) \in \Omega_1$ while the latter is satisfied by $e(0) \in \Omega_2$. The set $\Omega_1 \cap \Omega_2$ exists because it is always possible to select $F(0)$ sufficiently close to a framework in Iso(F^*).

The control (2.15) can be expressed element-wise as

$$u_i = -k_v \sum_{j \in \mathcal{N}_i(E^*)} \tilde{q}_{ij} z_{ij}, \quad i = 1, \ldots n, \tag{2.20}$$

which is only a function of $\tilde{q}_{ij}$ and d_{ij} for $(i,j) \in E^*$. Thus, the control law is *decentralized* in the sense of Definition 1.1 since it only requires the ith agent to measure its relative position to neighboring agents.

Notice that each individual term of the summation in (2.20) is a vector whose direction is along $\tilde{q}_{ij}$. If all n agents are positioned collinearly at $t = 0$ (see Figure 2.2a), the control input of each one will necessarily be directed along the line. As a result, the agents will be stuck in a collinear formation and will never converge to the desired formation. In other words, the collinear formation is an *invariant set*. However, if at least one agent is not initially collinear with the others (see Figure 2.2b), the agents will not necessarily remain collinear because the edges between these agents and the noncollinear ones will create control components whose directions are not parallel to the line.

Figure 2.2 (a) The collinearity of agents 1, 2, 3, and 4 is invariant. (b) The collinearity of agents 1, 3, and 4 is not invariant.

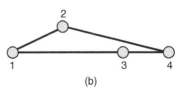

The stability result of Theorem 2.1 guarantees that the desired formation is acquired up to rotation and translation. In other words, the formation acquisition controller does not regulate the formation to a pre-defined global location in space. This is a reflection of the facts that u_i is not a function of q_i but only of the relative positions $\tilde{q}_{ij}$, $(i,j) \in E^*$ and that the control objective is to regulate $\|\tilde{q}_{ij}\|$.

Since we are only concerned with the inter-agent distances, any coordinate frame can be used to implement u_i. That is, although the above analysis was done with the variables defined with respect to a common, fixed coordinate frame for convenience, (2.20) can be implemented in practice with respect to the ith agent's own local coordinate frame. This means that the agents do not need to have a common sense of orientation and (2.20) is rotationally invariant. To see this, let $\mathcal{F}_0$ and $\mathcal{F}_i$ denote the Earth-fixed coordinate frame and the local coordinate frame of the ith agent, respectively (see Figure 2.3). If $\mathcal{R}_i^0 \in \mathbb{R}^m$ denotes the rotation matrix representing the orientation of $\mathcal{F}_i$ with respect to $\mathcal{F}_0$, we have that

$$\tilde{q}_{ij} := \tilde{q}_{ij}^0 = \mathcal{R}_i^0 \tilde{q}_{ij}^i$$

$$u_i := u_i^0 = \mathcal{R}_i^0 u_i^i$$

where the superscript denotes the coordinate frame in which the vector is specified. From (2.20), we can then write

$$u_i^i = -k_v \sum_{j \in \mathcal{N}_i(E^*)} (\mathcal{R}_i^0)^\mathsf{T} \tilde{q}_{ij} z_{ij}$$

$$= -k_v \sum_{j \in \mathcal{N}_i(E^*)} \tilde{q}_{ij}^i z_{ij}$$

since z_{ij} is independent of the coordinate frame.

Finally, the control (2.7) is in fact the standard *gradient descent law* that often appears in the literature (see, for example, [23]). If we rewrite z as

$$z = \phi(q) - \phi(q^*) \tag{2.21}$$

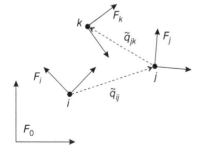

Figure 2.3 Fixed and local coordinate frames.

where (1.7) and (2.8) were used, it follows from (2.10) that

$$W = \frac{1}{4}\|\phi(q) - \phi(q^*)\|^2. \tag{2.22}$$

The derivative of (2.22) with respect to q is given by

$$\frac{\partial W}{\partial q} = \frac{1}{2}(\phi(q) - \phi(q^*))^\top \frac{\partial \phi(q)}{\partial q} = (\phi(q) - \phi(q^*))^\top R(\tilde{q})$$

where (1.15) was used. Therefore,

$$u = -\nabla_q W = -\left(\frac{\partial W}{\partial q}\right)^\top = -R^\top(\tilde{q})z,$$

which is the same as (2.7) without the control gain. That is, since (2.22) (also called a potential function) has a minimum when $\phi(q) = \phi(q^*)$, it is well known from optimization theory that the negative gradient causes the system trajectory to approach the local minimum.

2.2 Formation Maneuvering

In this section, we solve the formation maneuvering problem defined in Section 1.4 using model (2.1). Since formation acquisition is embedded in the formation maneuvering problem, we use (2.12) as the starting point. The control law here will take the form $u_i = u_i(\tilde{q}_{ij}, d_{ij}, v_{di})$, $i = 1, \ldots, n$ and $j \in \mathcal{N}_i(E^*)$ where $v_{di}(t)$, which was defined in (1.28), is a bounded continuous function.

Theorem 2.2 Consider the formation $F(t) = (G^*, q(t))$ with the initial conditions on $e(0)$ given in Theorem 2.1. Then, the control

$$u = u_a + v_d, \tag{2.23}$$

where u_a was defined in (2.15), $v_d = [v_{d1}, \ldots, v_{dn}] \in \mathbb{R}^{3n}$ is the desired rigid body velocity specified by[3]

$$v_{di} = v_0 + \omega_0 \times \tilde{q}_{in}, \quad i = 1, \ldots, n, \tag{2.24}$$

$v_0(t) \in \mathbb{R}^3$ denotes the desired translation velocity for the formation, $\omega_0(t) \in \mathbb{R}^3$ is the desired angular velocity, renders $e = 0$ exponentially stable and ensures that (1.26) and (1.28) are satisfied.

Proof: Substituting (2.23) into (2.12) yields

$$\dot{W} = -k_v z^T R(\tilde{q}) R^\top(\tilde{q}) z + z^T R(\tilde{q}) v_d. \tag{2.25}$$

---

3 Recall from the statement of the formation maneuvering problem in Section 1.4 that agent n serves as the reference point through which the rotation axis passes. Therefore, $\tilde{q}_{in}$ in (2.24) is the relative position between each agent and agent n.

It follows from (1.20) and (2.24) that

$$R(\tilde{q})v_d = 0. \tag{2.26}$$

Therefore, the proof of Theorem 2.1 can be directly followed to show that $e = 0$ is exponentially stable for $e(0) \in \Omega_1 \cap \Omega_2$ and (1.26) holds.

From (2.9) it is clear that $z \to 0$ as $e \to 0$. The exponential stability of $e = 0$ implies that $\tilde{q}$ is bounded from (2.6). Therefore, $R(\tilde{q})$ is bounded and we know from (2.23) and (2.15) that

$$u \to v_d \quad \text{as} \quad e \to 0. \tag{2.27}$$

Since we proved that $e(t) \to 0$ as $t \to \infty$, it follows from (2.27) and (2.1) that (1.28) holds. $\qquad\qquad\square$

The control (2.23) has two independent components: the term u_a is responsible for formation acquisition while v_d is responsible for rigid body maneuvers of the whole formation. We can see from (2.26) that the control exploits the special structure of the rigidity matrix to disassociate the formation acquisition stability analysis from the formation maneuvering analysis. Another interesting point is that, despite being based on the single-integrator model, (2.24) is generally not open-loop in nature since it depends on feedback of $\tilde{q}_{in}$. That is, (2.24) has an open-loop form only when the maneuver is purely translational.

The control law can be written element-wise as

$$u_i = -k_v \sum_{j \in \mathcal{N}_i(E^*)} \tilde{q}_{ij} z_{ij} + v_0 + \omega_0 \times \tilde{q}_{in}, \quad i = 1, \ldots n,$$

which shows that it is decentralized according to Definition 1.1. Note that in many applications the signals v_0 and ω_0 are known a priori and therefore can be stored on each agent's onboard computer. Also, since $\tilde{q}_{nn} = 0$, the formation maneuvering term of the leader only has the translation component v_0. This is expected since the leader by design lies on the axis of rotation of the virtual rigid body.

2.3 Flocking

Here, we consider the special case of formation maneuvering where the desired velocity only includes the translation component. Recall from Section 1.4 that this is commonly referred to as flocking. Unlike Section 2.2, we consider that the desired flocking velocity is only available to a subset of agents. We will overcome this constraint by employing a distributed observer that estimates this velocity by exploiting the connectedness of the formation graph.

2.3.1 Constant Flocking Velocity

We first consider the case where the flocking velocity is *constant*. Let $v_0 \in \mathbb{R}^m$ be the constant flocking velocity and $V_0 \subset V^*$ be the nonempty subset of agents that have direct access to v_0. To solve this flocking problem, we use the continuous controller–observer scheme

$$u = u_a + \hat{v} \tag{2.28a}$$

$$\dot{\hat{v}}_i = -\alpha \sum_{j \in \mathcal{N}_i(E^*)} (\hat{v}_i - \hat{v}_j) - \alpha b_i (\hat{v}_i - v_0), \quad i = 1, \dots n \tag{2.28b}$$

where

$$b_i = \begin{cases} 1, & \text{if } i \in V_0 \\ 0, & \text{otherwise,} \end{cases} \tag{2.29}$$

u_a was defined in (2.15), $\hat{v} = [\hat{v}_1, \dots, \hat{v}_n] \in \mathbb{R}^{mn}$ contains the velocity estimates for each agent, and $\alpha > 0$ is a user-defined observer gain.

Theorem 2.3 Consider the formation $F(t) = (G^*, q(t))$ with the initial conditions in Theorem 2.1. Then, the controller–observer scheme (2.28) with any $\hat{v}(0)$ renders $e = 0$ asymptotically stable and ensures that (1.26) and (1.28) are satisfied with $v_{di} = v_0, i = 1, \dots, n$.

Proof: Let

$$\tilde{v}_i = \hat{v}_i - v_0 \tag{2.30}$$

denote the flocking velocity estimation error for agent i. If $\tilde{v} = [\tilde{v}_1, \dots, \tilde{v}_n] \in \mathbb{R}^{mn}$, then

$$\tilde{v} = \hat{v} - \mathbf{1}_n \otimes v_0. \tag{2.31}$$

As part of this proof, we will show that (2.28b) guarantees $\tilde{v}(t) \to 0$ as $t \to \infty$. From the time derivative of (2.8), we have that

$$\dot{z} = 2R(\tilde{q})u. \tag{2.32}$$

After substituting (2.28a) into (2.32), we get the closed-loop system

$$\dot{z} = -2k_v R(\tilde{q}) R^\top (\tilde{q}) z + 2R(\tilde{q})\hat{v}. \tag{2.33}$$

Using (2.31) in (2.33) yields

$$\dot{z} = -2k_v R(\tilde{q}) R^\top (\tilde{q}) z + 2R(\tilde{q})\tilde{v} \tag{2.34}$$

upon application of Property 1.1.

Now, we turn our attention to deriving the dynamics of the estimation error. First, notice that

$$\sum_{j \in \mathcal{N}_i(E^*)} (\hat{v}_i - \hat{v}_j) = \sum_{j=1}^{n} a_{ij}(\hat{v}_i - \hat{v}_j)$$

where a_{ij} are the elements of the adjacency matrix defined in (1.3). Taking the time derivative of (2.31) and substituting (2.28b) gives

$$\dot{\tilde{v}} = -\alpha(\mathcal{L} \otimes I_m)\tilde{v} - \alpha(B \otimes I_m)\tilde{v}$$
$$= -\alpha(M \otimes I_m)\tilde{v} \qquad (2.35)$$

where we used the fact that $\hat{v}_i - \hat{v}_j = \tilde{v}_i - \tilde{v}_j$, $B := \text{diag}(b_1, \ldots b_n)$, $\mathcal{L}$ is the Laplacian matrix defined in (1.4), and $M := \mathcal{L} + B$ is symmetric. Our overall closed-loop system is composed of two interconnected subsystems, (2.34) and (2.35), which are in the form of (C.6). Notice that (2.34) with $\tilde{v} = 0$ is input-to-state stable by Theorem C.4 since it reduces to the closed-loop system analyzed in Theorem 2.1. Since the graph of a rigid framework is always connected, we know that G^* is connected. Therefore, we know from Lemmas 1.1 and C.1 that M and $M \otimes I_m$ are positive definite, respectively. It then follows from (2.35) that $\tilde{v} = 0$ is exponentially stable. We can now invoke Theorem C.5 to claim that $(z, \tilde{v}) = 0$ is an asymptotically stable equilibrium point of the interconnected system. Since $z = 0$ if and only if $e = 0$, we know $e = 0$ is asymptotically stable. Finally, by virtue of the initial conditions, we know that $F(t) \to \text{Iso}(F^*)$ as $t \to \infty$ as argued in the proof of Theorem 2.1.

Finally, due to the asymptotic stability of $e = 0$, we know $u_a(t) \to 0$ as $t \to \infty$ and therefore from (2.28a) that $u(t) - \hat{v}(t) \to 0$ as $t \to \infty$. Since $\tilde{v}_i(t) = \hat{v}_i(t) - v_0 \to 0$ as $t \to \infty$, then we know from (2.1) that (1.28) holds. □

The form of (2.28b) is inspired by multi-agent consensus algorithms [39]. The premise behind the observer is that agents that do not have direct access to v_0 can acquire this information from its neighbors since the graph modeling the communication network is connected. Note that the observer (2.28b) can accommodate a leader–follower strategy (only one agent has access to v_0) as well as the general case where the velocity information exchange happens between any two agents.

2.3.2 Time-Varying Flocking Velocity

The observer scheme in (2.28b) cannot be proven to ensure $\tilde{v}(t) \to 0$ as $t \to \infty$ for the case where the flocking velocity varies with time. In this situation, one can use the variable structure-type observer

$$\dot{\hat{v}}_i = -\alpha \text{sgn}\left(\sum_{j \in \mathcal{N}_i(E^*)} (\hat{v}_i - \hat{v}_j) + b_i(\hat{v}_i - v_0) \right), \quad i = 1, \ldots n \qquad (2.36)$$

where $v_0(t) \in \mathcal{L}_\infty$ is the time-varying flocking velocity, which is assumed to be differentiable with $\|\dot{v}_0(t)\|_{\mathcal{L}_\infty} \leq \gamma$ for all time, b_i was defined in (2.29), and $sgn(\cdot)$ is the standard signum function:

$$\text{sgn}(x) = \begin{cases} 1 & \text{for } x > 0 \\ 0 & \text{for } x = 0 \\ -1 & \text{for } x < 0. \end{cases} \tag{2.37}$$

The dynamics of the estimation error now become

$$\dot{\tilde{v}} = -\alpha \text{sgn}((M \otimes I_m)\tilde{v}) - \mathbf{1}_n \otimes \dot{v}_0 \tag{2.38}$$

where $\text{sgn}(x) = [sgn(x_1), \dots, sgn(x_n)]$, $\forall x \in \mathbb{R}^n$. Notice that (2.38) has a discontinuous right-hand side; thus, its solution needs to be studied using nonsmooth analysis (see Appendix C.5). Since $sgn(\cdot)$ is Lebesgue measurable and essentially locally bounded, one can show the existence of generalized solutions by embedding the differential equation into the differential inclusion

$$\dot{\tilde{v}} \in K[f](\tilde{v}, t) \tag{2.39}$$

where $K[\cdot]$ is a nonempty, compact, convex, upper semicontinuous set-valued map and $f(\tilde{v}, t) = -\alpha \text{sgn}((M \otimes I_m)\tilde{v}) - \mathbf{1}_n \otimes \dot{v}_0$.

If we define the Lyapunov function candidate

$$W_f = \frac{1}{2}\tilde{v}^\top (M \otimes I_m)\tilde{v}, \tag{2.40}$$

we get that [42]

$$\dot{W}_f \stackrel{a.e.}{\in} \frac{\partial W_f}{\partial \tilde{v}} K[f](\tilde{v}, t)$$

$$\subset -\alpha \tilde{v}^\top (M \otimes I_m)\text{sgn}((M \otimes I_m)\tilde{v}) - \tilde{v}^\top (M \otimes I_m)(\mathbf{1}_n \otimes \dot{v}_0) \tag{2.41}$$

where *a.e.* is the abbreviation for the term "almost everywhere". If we define $\text{SGN}(x) := [SGN(x_1), \dots, SGN(x_n)]$, $\forall x \in \mathbb{R}^n$ where

$$\text{SGN}(x_i) = \begin{cases} 1 & \text{for } x_i > 0 \\ [-1, 1] & \text{for } x_i = 0 \\ -1 & \text{for } x_i < 0, \end{cases} \tag{2.42}$$

then (2.41) becomes [42]

$$\dot{W}_f = -\alpha \tilde{v}^\top (M \otimes I_m)\text{SGN}((M \otimes I_m)\tilde{v}) - \tilde{v}^\top (M \otimes I_m)(\mathbf{1}_n \otimes \dot{v}_0)$$

$$= -\alpha \|(M \otimes I_m)\tilde{v}\|_1 - (\mathbf{1}_n \otimes \dot{v}_0)^\top (M \otimes I_m)\tilde{v}$$

$$= -\alpha \|(M \otimes I_m)\tilde{v}\|_1 + \dot{v}_0^\top \sum_{i=1}^{mn} [(M \otimes I_m)\tilde{v}]_i$$

$$\leq -\alpha \|(M \otimes I_m)\tilde{v}\|_1 + \|\dot{v}_0\|_1 \|(M \otimes I_m)\tilde{v}\|_1$$

$$\leq -(\alpha - \gamma)\|(M \otimes I_m)\tilde{v}\|_1. \tag{2.43}$$

By choosing $\alpha > \gamma$, we get that $\dot{W}_f$ is negative definite. Therefore, from Theorem C.6, we know that $\tilde{v} = 0$ is asymptotically stable.

Now the proof that (2.15) and (2.36) guarantee that (1.26) and (1.28) are satisfied directly follows from the proof of Theorem 2.3.

2.4 Target Interception with Unknown Target Velocity

We now turn our attention to the target interception problem defined in Section 1.4. We assume the target motion is such that $q_T(t)$ is three times continuously differentiable and $d^i q_T / dt^i \in \mathcal{L}_\infty$, $i = 0, 1, 2, 3$. Furthermore, we consider the target velocity $\dot{q}_T$ to be *unknown* to all agents, but that the leader can measure the target's relative position $q_T - q_n$ with its onboard sensors and can broadcast this information to the followers.

To simplify the notation, we let $v_T := \dot{q}_T$ and

$$e_T = q_T - q_n \tag{2.44}$$

denote the interception error between the leader and target. The control, which will include a term to "learn" the unknown target velocity, will take the general form $u_i = u_i(\tilde{q}_{ij}, d_{ij}, e_T, \hat{v}_T)$, $i = 1, \ldots, n$ and $j \in \mathcal{N}_i(E^*)$ where $\hat{v}_T$ is the target velocity estimate. This term is generated by the following continuous dynamic estimation mechanism

$$\hat{v}_T(t) = \int_0^t [k_1 e_T(\tau) + k_2 \mathrm{sgn}(e_T(\tau))]d\tau \tag{2.45}$$

where $k_1, k_2 > 0$ are user-defined control gains. This mechanism is inspired by the work in [24, 43] and allows one to learn or compensate for sufficiently smooth, nonperiodic signals.

Before presenting the control law, a lemma adopted from [24] that is related to (2.45) is introduced.

Lemma 2.2 Let

$$L := (k_1 e_T + \dot{e}_T)^\top (\dot{v}_T - k_2 \mathrm{sgn}(e_T)). \tag{2.46}$$

If k_2 in (2.45) is selected to satisfy the following sufficient condition

$$k_2 > \|\dot{v}_T\|_{\mathcal{L}_\infty} + \frac{1}{k_1}\|\ddot{v}_T\|_{\mathcal{L}_\infty}, \tag{2.47}$$

then

$$\int_0^t L(\tau)d\tau \leq \zeta_b \tag{2.48}$$

where the positive constant ζ_b is defined as

$$\zeta_b = k_2 \|e_T(0)\|_1 - e_T^{\mathsf{T}}(0)\dot{v}_T(0). \tag{2.49}$$

Proof: Integrating (2.46) over time yields

$$\int_0^t L(\tau)d\tau = \int_0^t (k_1 e_T(\tau) + \dot{e}_T(\tau))^{\mathsf{T}}[\dot{v}_T(\tau) - k_2 \operatorname{sgn}(e_T(\tau))]d\tau$$

$$= \int_0^t k_1 e_T^{\mathsf{T}}(\tau)[\dot{v}_T(\tau) - k_2 \operatorname{sgn}(e_T(\tau))]d\tau + \int_0^t \dot{e}_T^{\mathsf{T}}(\tau)\dot{v}_T(\tau)d\tau$$

$$- \int_0^t k_2 \dot{e}_T^{\mathsf{T}}(\tau)\operatorname{sgn}(e_T(\tau))d\tau. \tag{2.50}$$

After integrating by parts the second integral on the right-hand side of (2.50) and applying Lemma 1 of [44] to the third integral, we obtain

$$\int_0^t L(\tau)d\tau = \int_0^t k_1 e_T^{\mathsf{T}}(\tau)[\dot{v}_T(\tau) - k_2 \operatorname{sgn}(e_T(\tau))]d\tau$$

$$+ e_T^{\mathsf{T}}(\tau)\dot{v}_T(\tau)|_0^t - \int_0^t e_T^{\mathsf{T}}(\tau)\ddot{v}_T(\tau)d\tau - k_2 \|e_T(\tau)\|_1|_0^t$$

$$= \int_0^t k_1 e_T^{\mathsf{T}}(\tau)\left[\dot{v}_T(\tau) - \frac{1}{k_1}\ddot{v}_T(\tau) - k_2 \operatorname{sgn}(e_T(\tau))\right]d\tau$$

$$+ e_T^{\mathsf{T}}(t)\dot{v}_T(t) - e_T^{\mathsf{T}}(0)\dot{v}_T(0) - k_2 \|e_T(t)\|_1 + k_2 \|e_T(0)\|_1. \tag{2.51}$$

Using the fact that $\|x\|_1 \geq \|x\|$ for any $x \in \mathbb{R}^n$, we can upper bound the right-hand side of (2.51) by

$$\int_0^t L(\tau)d\tau \leq \int_0^t k_1 \|e_T(\tau)\| \left(\|\dot{v}_T(\tau)\| + \frac{1}{k_1}\|\ddot{v}_T(\tau)\| - k_2 \right)d\tau$$

$$+ \|e_T(t)\|(\|\dot{v}_T(t)\| - k_2) + k_2 \|e_T(0)\|_1 - e_T^{\mathsf{T}}(0)\dot{v}_T(0). \tag{2.52}$$

Applying (2.47) to (2.52) gives (2.48). Finally, the positiveness of (2.49) follows from the fact that

$$k_2 \|e_T(0)\|_1 - e_T^{\mathsf{T}}(0)\dot{v}_T(0) \geq \|e_T(0)\|(k_2 - \|\dot{v}_T(0)\|) > 0$$

when k_2 is selected according to (2.47). $\qquad\square$

Theorem 2.4 Consider the formation $F(t) = (G^*, q(t))$ with the initial conditions on $e(0)$ given in Theorem 2.1. Then, the control

$$u = u_a + \mathbf{1}_n \otimes h, \tag{2.53}$$

where $u_a = [u_{a1}, \ldots, u_{an}]$ was defined in (2.15) and

$$h = (k_1 + 1)e_T + \hat{v}_T - u_{an}, \tag{2.54}$$

renders $e = 0$ exponentially stable and ensures that (1.26) and (1.29) are satisfied. Further, the target velocity can be identified in the sense that $v_T(t) - \hat{v}_T(t) \to 0$ as $t \to \infty$.

Proof: After substituting (2.53) into (2.12), we obtain

$$\dot{W} = -k_v z^\top R(\tilde{q}) R^\top(\tilde{q}) z + z^\top R(\tilde{q})(\mathbf{1}_n \otimes h). \tag{2.55}$$

Due to Property 1.1, the second term on the right-hand side of (2.55) disappears and the proof of Theorem 2.1 can be again followed to prove the exponential stability of $e = 0$ and (1.26).

We now proceed to prove (1.29). From (2.53) and (2.54), we have that the leader control input is[4]

$$u_n = (k_1 + 1)e_T + \hat{v}_T. \tag{2.56}$$

Differentiating (2.44) and using (2.56) yields

$$\dot{e}_T = v_T - u_n \tag{2.57}$$

$$= v_T - (k_1 + 1)e_T - \hat{v}_T \tag{2.58}$$

$$= -k_1 e_T + w \tag{2.59}$$

where

$$w = v_T - e_T - \hat{v}_T. \tag{2.60}$$

The derivative of (2.60) is given by

$$\dot{w} = \dot{v}_T - \dot{e}_T - k_1 e_T - k_2 \mathrm{sgn}(e_T) = -w + \dot{v}_T - k_2 \mathrm{sgn}(e_T) \tag{2.61}$$

where (2.45) and (2.59) were used.

Next, define the auxiliary function

$$P = \frac{1}{2} w^\top w, \tag{2.62}$$

whose derivative along (2.61) is given by

$$\dot{P} = w^\top(-w + \dot{v}_T - k_2 \mathrm{sgn}(e_T)) = -w^\top w + L \tag{2.63}$$

where (2.46) was used. After integrating both sides of (2.63) with respect to time and applying Lemma 2.2, we obtain

$$\int_0^t \dot{P}(\tau)d\tau = P(t) - P(0) = -\int_0^t w^\top(\tau)w(\tau)d\tau + \int_0^t L(\tau)d\tau$$

$$\leq -\int_0^t w^\top(\tau)w(\tau)d\tau + \zeta_b \leq \zeta_b. \tag{2.64}$$

---

4 The introduction of the term $-u_{an}$ in (2.54) is crucial for the following stability analysis of the target interception error since it allows u_n to have the simple form in (2.56).

Since $P(0)$ is finite, it follows from (2.64) that $P(t) \in \mathcal{L}_\infty$, which implies that $w(t) \in \mathcal{L}_\infty$ from (2.62). From (2.64), we also have that

$$\int_0^t w^\top(\tau)w(\tau)d\tau \leq \zeta_b + P(0) - P(t) < \infty,$$

which means that $w(t) \in \mathcal{L}_2$. Therefore, we know from (2.59) and Theorem C.1 that $e_T(t) \to 0$ as $t \to \infty$. We can also use (2.59) to claim that $\dot{e}_T \in \mathcal{L}_\infty$, which implies from (2.57) (together with the boundedness of $v_T(t)$) that $u_n(t) \in \mathcal{L}_\infty$. From (2.56), we then know that $\hat{v}_T(t) \in \mathcal{L}_\infty$. Since (1.26) holds and F^* is constructed such that $q_n^* \in \mathrm{conv}\{q_1^*, ..., q_{n-1}^*\}$, we know that $q_n(t) \in \mathrm{conv}\{q_1(t), q_2(t), ..., q_{n-1}(t)\}$ as $t \to \infty$. Therefore, from the fact that $e_T(t) \to 0$ as $t \to \infty$, we conclude that (1.29) holds.

Finally, we know $\dot{w}(t) \in \mathcal{L}_\infty$ from (2.61) since $\dot{v}_T$ is assumed bounded. It then follows from Theorem C.3 that $w(t) \to 0$ as $t \to \infty$. Therefore, we can use (2.59) to show that $\dot{e}_T(t) \to 0$ as $t \to \infty$, and then (2.58) to conclude that $v_T(t) - \hat{v}_T(t) \to 0$ as $t \to \infty$. $\qquad\square$

Similar to the formation maneuvering control, the target interception controller (2.53) and (2.54) has two components with well-defined roles: u_a ensures formation acquisition while the term h guarantees target interception. The controller for the followers can be written element-wise as

$$u_i = -k_v \sum_{j \in \mathcal{N}_i(E^*)} \tilde{q}_{ij} z_{ij} + (k_1 + 1)e_T + \int_0^t [k_1 e_T(\tau) + k_2 \mathrm{sgn}(e_T(\tau))]d\tau - u_{an},$$

for $i = 1, ..., n-1$ where

$$u_{an} = -k_v \sum_{j \in \mathcal{N}_n(E^*)} \tilde{q}_{nj} z_{nj},$$

whereas the control for the leader is given by (2.56). As one can see, each follower control input depends on its relative position to neighboring agents, the target interception error, and the formation acquisition control term of the leader. Therefore, it is less decentralized than the formation acquisition and maneuvering controllers since now information needs to be wirelessly broadcast from the leader to the followers, but still in conformance with Definition 1.1.

Finally, note that the target interception error (2.44) could be redefined to include a constant offset so that the leader does not collide with the target, i.e., $e_T = q_n - q_T - c$ where $c \in \mathbb{R}^m$ is a constant vector.

2.5 Dynamic Formation Acquisition

So far, we have only considered formation acquisition when the desired formation F^* is static. In certain applications it may be necessary that the formation size and/or geometric shape change in time, such as to avoid obstacles, dynamically adapt to a change of mission, or adapt to limits in communication range and bandwidth. Thus, we consider now the problem of *dynamic* formation acquisition in the sense that the desired formation is a function of time, $F^*(t)$. In control systems jargon, we will deal here with the more general *tracking* problem instead of the simpler setpoint problem.

Note that dynamic formation acquisition is independent of what we call formation maneuvering. In the former, the time-varying nature is related to the formation itself, whereas in the latter, the formation (whether static or dynamic) maneuvers as a virtual rigid body. The formal statement of the dynamic formation acquisition problem is as follows.

Problem 1 *(Dynamic Formation Acquisition)* Let the desired formation be represented by a *dynamic*, infinitesimally and minimally rigid framework $F^*(t) = (G^*, q^*(t))$[5] where the time-varying desired distance between agents i and j is given by

$$d_{ij}(t) = \|q_i^*(t) - q_j^*(t)\| > 0, \quad i,j \in V^*. \tag{2.65}$$

We assume the desired distances are sufficiently smooth functions of time.[6] The control objective is to design u_i such that

$$F(t) - \mathrm{Iso}(F^*(t)) \to 0 \text{ as } t \to \infty, \tag{2.66}$$

or equivalently

$$e_{ij}(t) \to 0 \text{ as } t \to \infty, \quad i,j \in V^*. \tag{2.67}$$

$\triangle$

Because of the time-varying nature of (2.65), the distance error dynamics is now given by

$$\dot{e}_{ij} = \frac{\tilde{q}_{ij}^{\mathsf{T}}(u_i - u_j)}{e_{ij} + d_{ij}} - \dot{d}_{ij}, \tag{2.68}$$

where (2.6) and (2.1) were used. As a result, the derivative of (2.10) along (2.68) becomes

$$\dot{W} = \sum_{(i,j) \in E^*} e_{ij}(e_{ij} + 2d_{ij})[\tilde{q}_{ij}^{\mathsf{T}}(u_i - u_j) - d_{ij}\dot{d}_{ij}] = z^{\mathsf{T}}(R(\tilde{q})u - d_v) \tag{2.69}$$

where

$$d_v = [..., d_{ij}\dot{d}_{ij}, ...] \in \mathbb{R}^l, \quad (i,j) \in E^* \tag{2.70}$$

---

5 It is important to point out that the framework $F^*(t)$ is required to be infinitesimally and minimally rigid for all time.

6 Since the precise smoothness properties are agent model-dependent, they will be specified later.

with elements ordered as (1.7). We assume d_{ij} is a continuously differentiable function of time and $d_{ij}(t), \dot{d}_{ij}(t) \in \mathcal{L}_\infty$. The presence of the extra term, d_v, in the derivative of the Lyapunov function candidate will dictate a different control structure.

Theorem 2.5 Consider the formation $F(t) = (G^*, q(t))$ with the initial conditions given in Theorem 2.1. The control law

$$u = R^+(\tilde{q})(-k_v z + d_v), \tag{2.71}$$

where $R^+(\tilde{q}) = R^\top(\tilde{q})[R(\tilde{q})R^\top(\tilde{q})]^{-1}$ is the Moore–Penrose pseudoinverse, yields $e = 0$ exponentially stable and guarantees that (2.66) is satisfied.

The proof of this theorem is nearly identical to the proof of Theorem 2.1 so the details are omitted. The main difference is that, since $R(\tilde{q})$ has full row rank for $e(t) \in \Omega_1$, then $R(\tilde{q})R^+(\tilde{q}) = I$ [45] for $e(t) \in \Omega_1$. Therefore, substituting (2.71) into (2.69) yields

$$\dot{W} = -k_v z^\top z = -4k_v W \quad \text{for } e(t) \in \Omega_1. \tag{2.72}$$

From this point on, the proof of Theorem 2.1 can be directly followed to show that (2.66) holds for $e(0) \in \Omega_1 \cap \Omega_2$.

A fundamental difference exists in the implementation of (2.71) in comparison to the previous controllers of this chapter. Namely, the matrix $R^+(\tilde{q})$ couples the variables such that $u_i = u_i(\tilde{q}_{ij}, d_{ij}, \dot{d}_{ij})$, $i = 1, \ldots, n$ and $(i, j) \in E^*$. That is, unlike in the previous cases where $j \in \mathcal{N}_i(E^*)$ for the ith input, here each input is dependent on *all* $(i, j) \in E^*$ variables.

Formation maneuvering can be performed on top of dynamic formation acquisition by modifying (2.71) to

$$u = R^+(\tilde{q})(-k_v z + d_v) + v_d \tag{2.73}$$

where v_d was defined in (2.24). It is straightforward to show that (2.73) ensures (1.28) by following the proof of Theorem 2.2.

2.6 Simulation Results

In this section, simulations are presented to demonstrate the performance of the above-designed formation controllers. The simulations show agents moving in two-dimensional (2D) and three-dimensional (3D) space. The simulations were carried out in MATLAB using ODE solver ode45.

2.6.1 Formation Acquisition

A simulation of five agents was conducted to show that control objective (1.26) is achieved by applying control input (2.15) to (2.1). The desired formation F^*

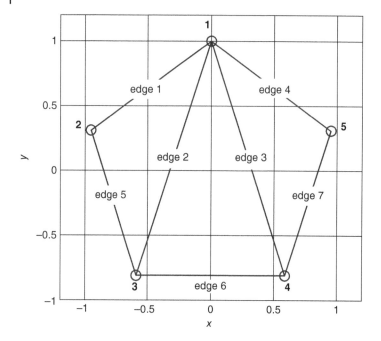

Figure 2.4 Formation acquisition: desired formation F^*.

was set to the regular convex pentagon shown in Figure 2.4 with vertices at coordinates $[0, 1]$, $[-s_1, c_1]$, $[-s_2, -c_2]$, $[s_2, -c_2]$, $[s_1, c_1]$ where

$$s_1 = \sin 2\pi/5, \quad s_2 = \sin 4\pi/5, \quad c_1 = \cos 2\pi/5, \quad \text{and} \quad c_2 = \cos \pi/5. \tag{2.74}$$

The vertices of the framework were ordered counterclockwise with the coordinate $[0, 1]$ as vertex number 1 (i.e., $q_1^* = [0, 1]$). The desired framework was made minimally rigid and infinitesimally rigid by introducing seven edges and leaving the vertex pairs $(2, 5)$, $(2, 4)$, and $(3, 5)$ disconnected. That is,

$$E^* = \{(1, 2), (1, 3), (1, 4), (1, 5), (2, 3), (3, 4), (4, 5)\}. \tag{2.75}$$

Thus, the desired distances being directly controlled were $d_{12} = d_{15} = d_{23} = d_{34} = d_{45} = \sqrt{2(1 - c_1)}$ and $d_{13} = d_{14} = \sqrt{2(1 + c_2)}$.

The initial conditions of the agents were randomly set by

$$q_i(0) = q_i^* + \delta[\text{rand}(0, 1) - 0.5\mathbf{1}_2] \tag{2.76}$$

where $i = 1, \ldots, 5$, $\delta = 1$, and $\text{rand}(0, 1)$ generates a random 2×1 vector whose elements are uniformly distributed on the interval $(0, 1)$. The control gain k_v in (2.15) was set to 1. Figure 2.5 shows the trajectories of the five agents forming

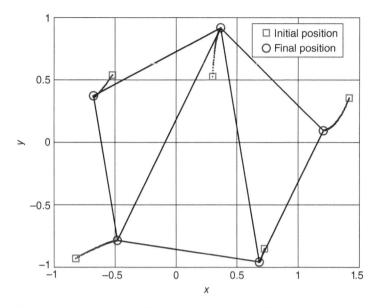

Figure 2.5 Formation acquisition: agent trajectories $q_i(t), i = 1, ..., 5$.

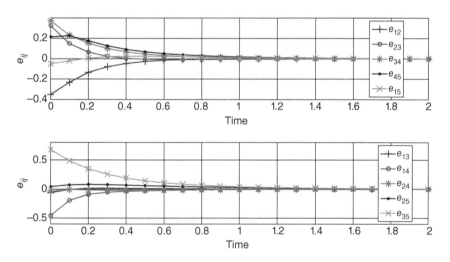

Figure 2.6 Formation acquisition: distance errors $e_{ij}(t), i, j \in V^*$.

the desired shape (up to rotation and translation), while Figure 2.6 shows the distance errors $e_{ij}(t), i, j \in V^*$ approaching zero[7]. The x- and y-direction components of the control inputs $u_i(t), i = 1, ..., 5$ are given in Figure 2.7.

---

7 The reason why *ten* errors are shown in Figure 2.6 is because $n(n - 1)/2 = 10$ when $n = 5$.

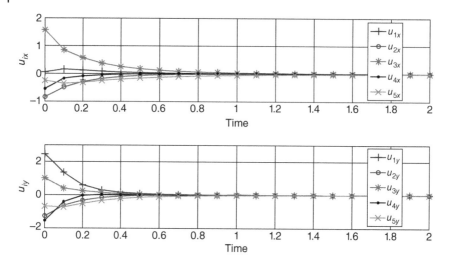

Figure 2.7 Formation acquisition: control inputs $u_i(t)$, $i = 1, \ldots, 5$.

A second simulation was conducted to illustrate the local nature of the stability result, as discussed in Section 2.1. To this end, the distance between each agent at $t = 0$ and its corresponding target position was increased by selecting the initial conditions according to (2.76) with $\delta = 2$. The simulation results in Figure 2.8 show the agents reaching an incorrect formation due to $e(0) \notin \Omega_1 \cap \Omega_2$. Notice that this incorrect formation corresponds to a flip ambiguity where edges 4 and 7 are flipped over edge 3 (see labels in Figure 2.4). That is, in this case, the control drives the agents to a formation with guaranteed desired inter-agent distances *only* for agent pairs $(i, j) \in E^*$ since the agents are initially positioned such that $\text{dist}(q(0), \text{Iso}(F^*)) > \text{dist}(q(0), \text{Amb}(F^*))$. This is clear from the plots in Figure 2.9 where the distance errors have been segregated by whether $(i, j) \in E^*$ (subplot (a)) or $(i, j) \notin E^*$ (subplot (b)). One can see that all $e_{ij}(t)$ with $(i, j) \in E^*$ converge to zero, but the $e_{ij}(t)$ values that are not in the edge set (2.75) do not necessarily converge to zero.

Figure 2.8 indicates that agent 5 is the one that is not properly initialized and therefore causing convergence to the incorrect formation. Now consider that we modify the edge set of the framework in Figure 2.4 to

$$E^* = \{(1, 2), (2, 3), (3, 4), (4, 5), (1, 5), (2, 4), (2, 5)\},$$

leading to the framework in Figure 2.10. Notice that agent 5 is now connected to agent 2, ensuring that this distance is directly controlled. If the simulation is re-run with the exact same initial conditions as Figure 2.8, the agents will converge to the desired formation (up to rotation and translation), as shown in Figure 2.11. This suggests that, in practice, the edges of the desired framework should be judiciously chosen based on the actual formation at $t = 0$ to avoid convergence to an ambiguous framework.

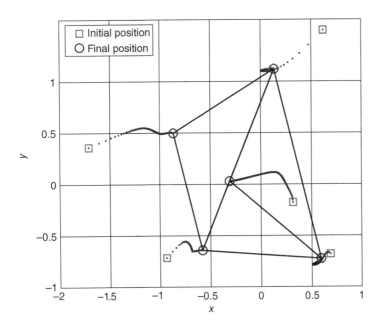

Figure 2.8 Formation acquisition: agents converging to an ambiguous framework.

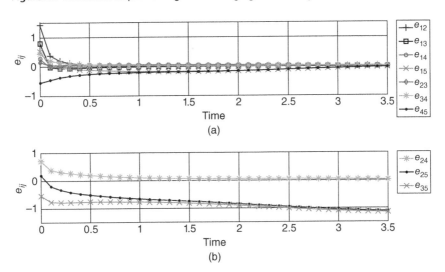

Figure 2.9 Formation acquisition: (a) distance errors $e_{ij}(t)$, $(i,j) \in E^*$ and (b) distance errors $e_{ij}(t)$, $(i,j) \notin E^*$.

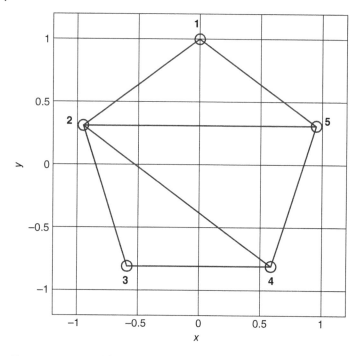

Figure 2.10 Desired formation F^* with modified edge set.

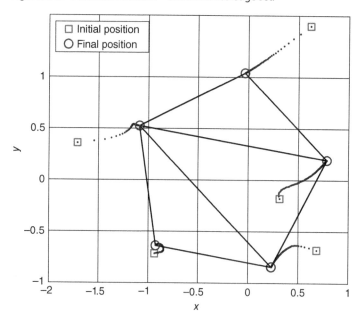

Figure 2.11 Formation acquisition: agents converging to correct formation after change in edge set.

2.6.2 Formation Maneuvering

The purpose here was to simulate the use of the control law given by (2.23), (2.15), and (2.24) to achieve (1.26) and (1.28). Two simulations were conducted: one in 2D and the other in 3D.

Simulation 1: In this simulation, the agents were required to form a regular convex pentagon on the plane while maneuvering according to (2.24) with

$$v_0 = [1, \cos t, 0] \quad \text{and} \quad \omega_0 = [0, 0, 1]. \tag{2.77}$$

The desired formation F^* is shown in Figure 2.12. The five outer vertices of the pentagon represent the followers while the vertex at the geometric center of the pentagon represents the leader about which the followers will rotate. Thus, here, $V^* = \{1, 2, 3, 4, 5, 6\}$. The desired framework was made minimally rigid and infinitesimally rigid by introducing $9 \ (= 2 \times 6 - 3)$ edges with all followers connected to the leader. The edge set was selected as

$$E^* = \{(1, 2), (1, 6), (2, 3), (2, 6), (3, 4), (3, 6), (4, 5), (4, 6), (5, 6)\}.$$

The desired inter-agent distances were $d_{12} = d_{23} = d_{34} = d_{45} = \sqrt{2(1 - c_1)}$ and $d_{16} = d_{26} = d_{36} = d_{46} = d_{56} = 1$.

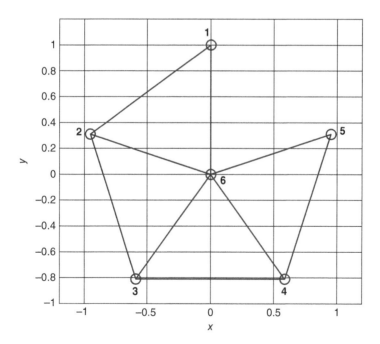

Figure 2.12 Formation maneuvering (Simulation 1): desired formation F^*.

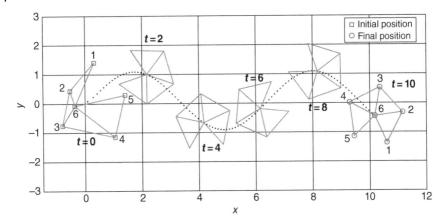

Figure 2.13 Formation maneuvering (Simulation 1): snapshots of $F(t)$ at different instants of time.

The initial conditions of the agents were selected by (2.76) with $i = 1, \dots, 6$ and $\delta = 1$, while the control gain k_v was set to 1.

Figure 2.13 shows snapshots of the actual formation $F(t)$ over time, where the dotted line represents the leader trajectory. Notice that by $t = 2$, the agents have virtually formed the desired shape while simultaneously translating and rotating according to (2.77). The rotation is counterclockwise about the axis coming out of the page based on the right-hand rule. All the distance errors $e_{ij}(t)$, $i, j \in V^*$ are depicted in Figure 2.14 and converge to zero as expected. The x- and y-direction components of the control inputs $u_i(t)$, $i = 1, \dots, 6$ are given in Figure 2.15. Notice that in the x (resp., y) direction, the control input converges to the first (resp., second) element of v_0 in (2.77) plus the contribution from the angular velocity (see (2.24)).

Simulation 2: Here, the desired formation was selected to be a cube with edge length of 2 as shown in Figure 2.16. For ease of implementation, the eight vertices of the cube were chosen to represent the followers while the geometric center of the cube (point $[0, 0, 0]$) was chosen to represent the leader (i.e., agent 9), thus $V^* = \{1, \dots, 9\}$. The required minimally rigid and infinitesimally rigid conditions of F^* were enforced by introducing 21 ($= 3 \times 9 - 6$) edges with all followers connected to the leader. The edge set was set to

$$E^* = \{(1, 4), (1, 5), (1, 9), (2, 3), (2, 6), (2, 9), (3, 4), (3, 7), (3, 9), (4, 5), (4, 8),$$
$$(4, 9), (5, 6), (5, 8), (5, 9), (6, 7), (6, 8), (6, 9), (7, 8), (7, 9), (8, 9)\}. \quad (2.78)$$

The desired inter-agent distances were given by $d_{14} = d_{15} = d_{23} = d_{26} = d_{34} = d_{37} = d_{48} = d_{56} = d_{58} = d_{67} = d_{78} = 2$, $d_{19} = d_{29} = d_{39} = d_{49} = d_{59} = d_{69} = d_{79} = d_{89} = \sqrt{3}$, and $d_{45} = d_{68} = 2\sqrt{2}$. The initial conditions were chosen as in Simulation 1.

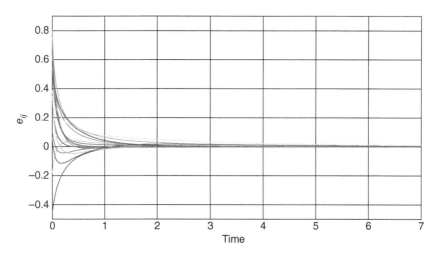

Figure 2.14 Formation maneuvering (Simulation 1): distance errors $e_{ij}(t)$, $i, j \in V^*$.

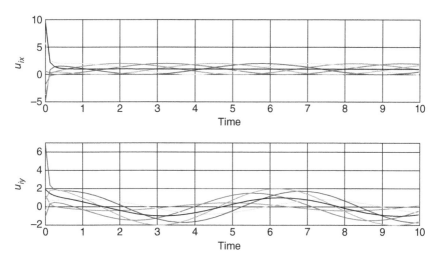

Figure 2.15 Formation maneuvering (Simulation 1): control inputs $u_i(t)$, $i = 1, \ldots, 6$.

The maneuvering velocity was set to

$$v_0 = [5 \cos \pi t/3, 5 \sin \pi t/3, 0] \quad \text{and} \quad \omega_0 = [0, 0, \pi/3], \tag{2.79}$$

which corresponds to a circular orbit of the formation about the z-axis with the same side of the cube always facing the interior of the orbit. The control gain was again chosen as 1. Figure 2.17 shows snapshots of the formation maneuver, the distance errors are shown in Figure 2.18, and the control inputs are shown in Figure 2.19.

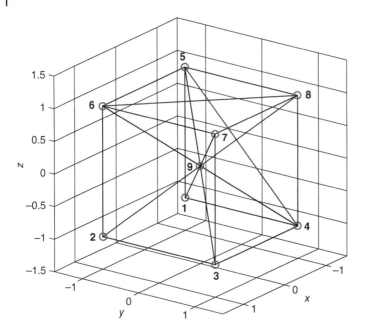

Figure 2.16 Formation maneuvering (Simulation 2): desired formation F^*.

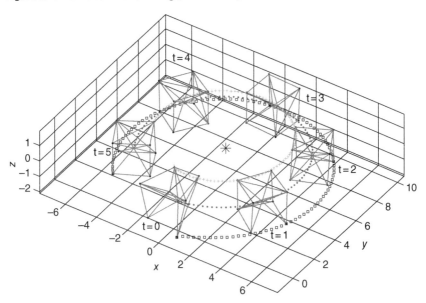

Figure 2.17 Formation maneuvering (Simulation 2): snapshots in time of the formation maneuver. The asterisk denotes the orbit center, the dots are the leader trajectory, the squares are the agent 2 trajectory, and the crosses are the agent 8 trajectory.

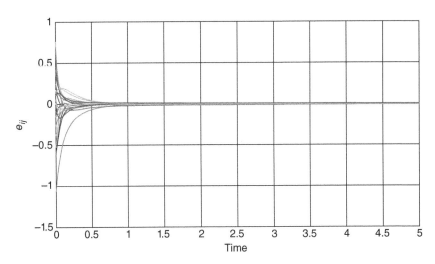

Figure 2.18 Formation maneuvering (Simulation 2): distance errors $e_{ij}(t)$, $(i,j) \in E^*$.

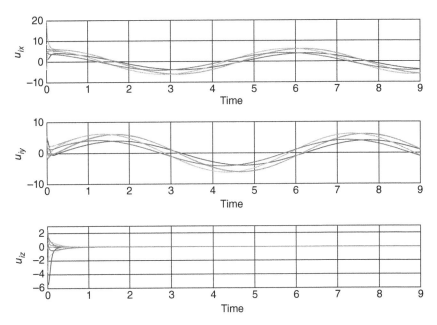

Figure 2.19 Formation maneuvering (Simulation 2): control inputs $u_i(t)$, $i = 1, \ldots 9$ in the x, y, and z directions.

2.6.3 Flocking

The flocking schemes from Section 2.3 were simulated for the case where five agents have to acquire the desired formation shown in Figure 2.4 and flock with some desired velocity. The control and observer gains in (2.28) and (2.36) were set to $k_v = 1$ and $\alpha = 10$, respectively. The initial positions of the agents were randomly generated by (2.76) with $\delta = 1$ while $\hat{v}(0) = 0$.

First, the controller–observer scheme in (2.28) was simulated with desired velocity $v_0 = [1, 2]$ under two conditions: one where only agent 2 has knowledge of v_0 ($V_0 = \{2\}$) and the other where $V_0 = \{2, 3\}$. The trajectory of the actual formation $F(t)$ is shown in Figure 2.20 for the case where $V_0 = \{2\}$. The inter-agent distances errors and velocity estimation errors are shown Figures 2.21 and 2.22, respectively. The control inputs are given in Figure 2.23. As can be seen from the figures, the agents meet the flocking objective by approximately $t = 3$. Notice from Figure 2.22 that the estimation error with the fastest convergence is $\tilde{v}_2(t)$ since agent 2 has direct access to v_0. This

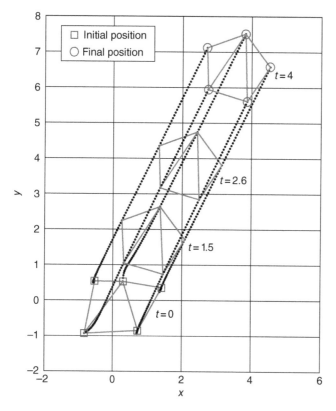

Figure 2.20 Flocking (continuous controller–observer; $V_0 = \{2\}$): snapshots of $F(t)$ at different instants of time.

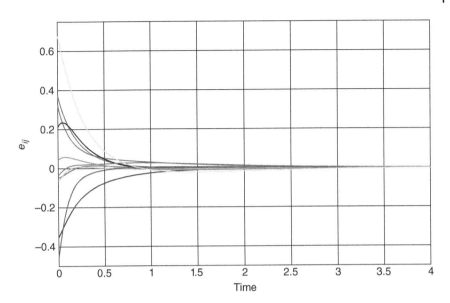

Figure 2.21 Flocking (continuous controller–observer; $V_0 = \{2\}$): distance errors $e_{ij}(t)$, $i, j \in V^*$.

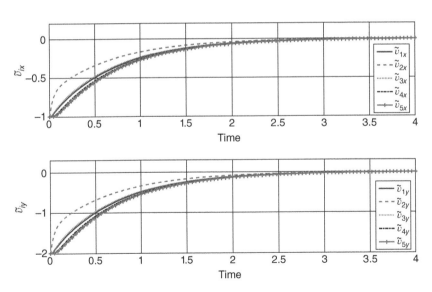

Figure 2.22 Flocking (continuous controller–observer; $V_0 = \{2\}$): velocity estimation errors $\tilde{v}_i(t)$, $i = 1, \ldots 5$ in the x and y directions.

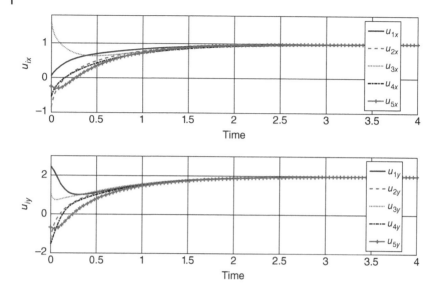

Figure 2.23 Flocking (continuous controller–observer; $V_0 = \{2\}$): control inputs $u_i(t)$, $i = 1, \ldots 5$ in the x and y directions.

is followed by $\tilde{v}_1(t)$ and $\tilde{v}_3(t)$, which are directly connected to agent 2 (see Figure 2.4), and then $\tilde{v}_4(t)$ and $\tilde{v}_5(t)$, which are one hop away from agent 2.

The results for the case where $V_0 = \{2, 3\}$ are given in Figures 2.24–2.26. One can see from Figure 2.25 that the estimation errors of all agents converge faster than in Figure 2.22 since the flocking velocity is available to more agents, resulting in faster consensus. This is also reflected in the distance errors, which also have a faster convergence.

Next, the controller–observer in (2.28) was simulated with desired velocity $v_0(t) = [1, \cos t]$ and $V_0 = \{2\}$. Due to the inability of observer (2.27) in ensuring $\tilde{v}(t) \to 0$ as $t \to \infty$ when v_0 is time varying, one can see from Figure 2.27 that the flocking velocity estimation error in the y direction is nonzero. This also affects the formation acquisition as evidenced by the inter-agent distances errors in Figure 2.28.

When the discontinuous controller–observer (2.28a) and (2.36) is applied to the same time-varying velocity case, we get the results shown in Figures 2.29–2.31. The variable structure nature of the observer allows for a fast estimation of the flocking velocity, which in turn causes the agent velocities to converge to $v_0(t)$ by $t = 1$.

2.6.4 Target Interception

In this simulation, we tested the control law defined by (2.53), (2.54), and (2.45) in satisfying (1.26) and (1.29). The unknown velocity of the moving target was

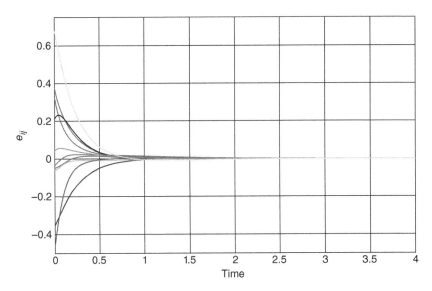

Figure 2.24 Flocking (continuous controller–observer; $V_0 = \{2, 3\}$): distance errors $e_{ij}(t)$, $i, j \in V^*$.

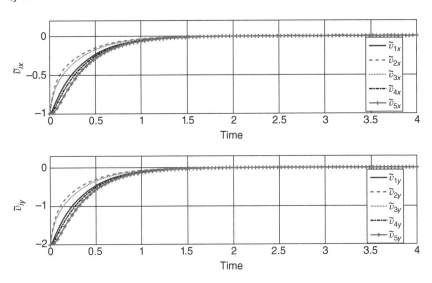

Figure 2.25 Flocking (continuous controller–observer; $V_0 = \{2, 3\}$): velocity estimation errors $\tilde{v}_i(t)$, $i = 1, \ldots 5$ in the x and y directions.

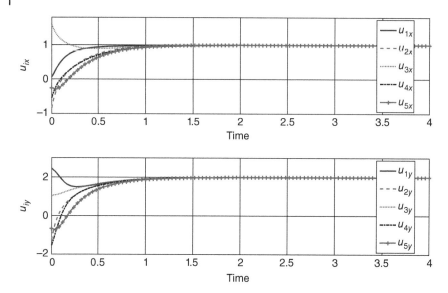

Figure 2.26 Flocking (continuous controller–observer; $V_0 = \{2, 3\}$): control inputs $u_i(t)$, $i = 1, \ldots 5$ in the x and y directions.

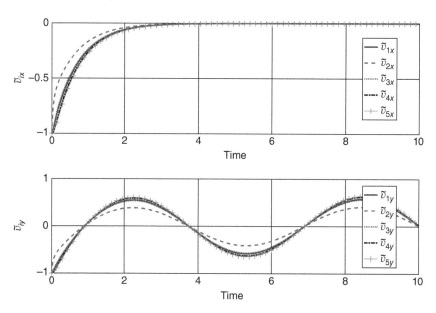

Figure 2.27 Flocking (continuous controller–observer; $V_0 = \{2\}$): velocity estimation errors with time-varying desired velocity.

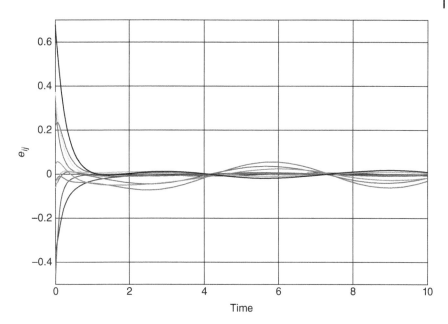

Figure 2.28 Flocking (continuous controller–observer; $V_0 = \{2\}$): distance errors with time-varying desired velocity.

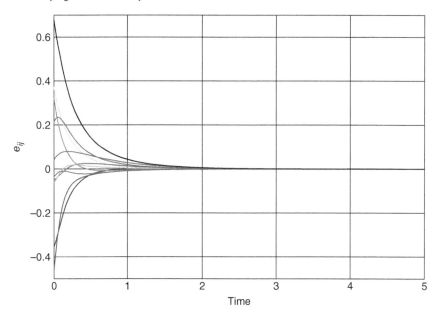

Figure 2.29 Flocking (discontinuous controller–observer; $V_0 = \{2\}$): distance errors with time-varying desired velocity.

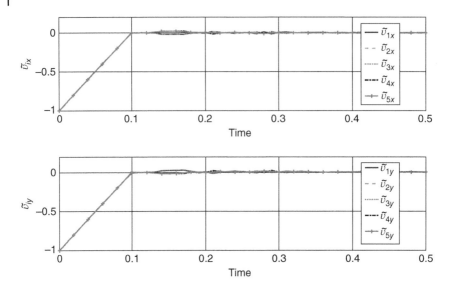

Figure 2.30 Flocking (discontinuous controller–observer; $V_0 = \{2\}$): velocity estimation errors with time-varying desired velocity.

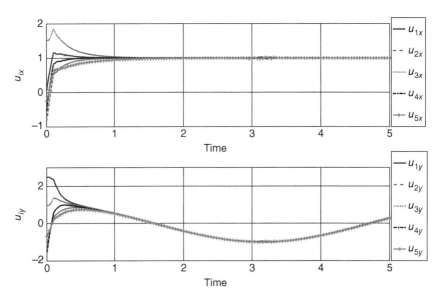

Figure 2.31 Flocking (discontinuous controller–observer; $V_0 = \{2\}$): control inputs.

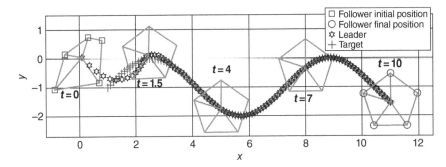

Figure 2.32 Target interception: snapshots of $F(t)$ at different instants of time along with target motion.

set to

$$v_T = [1, \cos t] \qquad (2.80)$$

with initial position $q_T(0) = [1, -1]$. We used the same desired formation topology defined in Section 2.6.2 (see Figure 2.12) with agent 6 being the leader that is responsible for tracking the target. The initial conditions were again set by (2.76) with $i = 1, ..., 6$ and $\delta = 1$ while $\hat{v}_T(0) = 0$. The control gains were set to $k_v = 1$, $k_1 = 2$, and $k_2 = 2$. The fact that this value for k_2 does not satisfy (2.47) since $\|\dot{v}_T\|_{\mathcal{L}_\infty} = \|\ddot{v}_T\|_{\mathcal{L}_\infty} = \sqrt{2}$ from (2.80) will demonstrate that this condition is indeed only sufficient for stability of the target tracking error.

Figure 2.32 shows the leader intercepting the target and the followers simultaneously enclosing it with the pentagon formation. The inter-agent distance errors are given in Figure 2.33. Figure 2.34 displays the control inputs of each agent converging to v_T (viz., 1 in the x direction and $\cos t$ in the y direction), despite the target velocity being unknown.

2.6.5 Dynamic Formation

We simulated a scenario where the formation needs to vary in order to move through a narrow passageway. To this end, control law (2.73) was simulated since it includes the dynamic formation acquisition control law (2.71). The primary objective here is (2.66).

The desired dynamic formation $F^*(t)$ was set to a regular convex pentagon that expanded and contracted uniformly over time.[8] The desired formation at $t = 0$, $F^*(0)$, is the same one used in the simulation of Section 2.6.1 and depicted in Figure 2.4. The desired formation was made dynamic by setting the vertex

---

8 In practice, $F^*(t)$ would be generated on-line by a path planning module that includes a perceptive model of the environment.

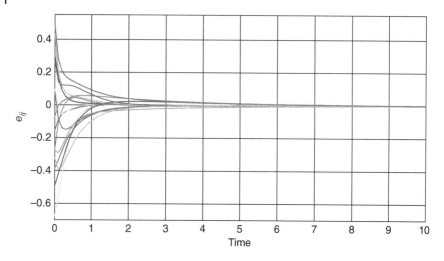

Figure 2.33 Target interception: distance errors $e_{ij}(t)$, $i, j \in V^*$.

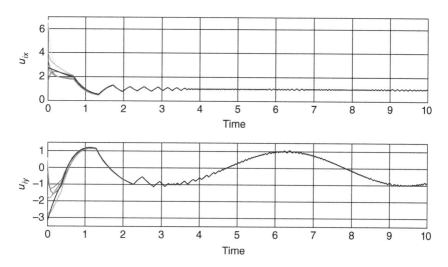

Figure 2.34 Target interception: control inputs $u_i(t)$, $i = 1, \ldots, 6$.

coordinates to

$$q_1^*(t) = [0, 1]\rho(t), \quad q_2^*(t) = [-s_1, c_1]\rho(t), \quad q_3^*(t) = [-s_2, -c_2]\rho(t),$$
$$q_4^*(t) = [s_2, -c_2]\rho(t), \quad q_5^*(t) = [s_1, c_1]\rho(t)$$

where (2.74) was used and

$$\rho(t) = 1 + 0.5 \sin 0.4t. \tag{2.81}$$

As a result, the desired inter-agent distances were given by

$$d_{12}(t) = d_{15}(t) = d_{23}(t) = d_{34}(t) = d_{45}(t) = \rho(t)\sqrt{2(1 - c_1)}$$
$$d_{13}(t) = d_{14}(t) = \rho(t)\sqrt{2(1 + c_2)}.$$

The desired swarm velocity was set to $v_d = [1, \cos t]$. The initial conditions of the agents were selected by (2.76) with $i = 1, \ldots, 5$ and $\delta = 1$. The control gain k_v was set to 10.

Figure 2.35 shows the agent trajectories over time as they track the desired formation while translating according to v_d. The effective tracking of $F^*(t)$ is better depicted by Figure 2.36, which shows the inter-agent distance errors approaching zero. The x- and y-direction components of the control inputs are given in Figure 2.37, where the left plots show their transient behavior ($0 \leq t \leq 0.5$) and the right plots show the steady-state behavior ($0.5 \leq t \leq 15$).

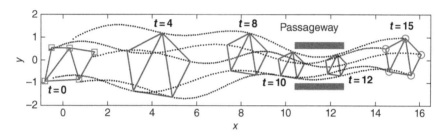

Figure 2.35 Dynamic formation: snapshots of $F(t)$ at different instants of time along with agent trajectories.

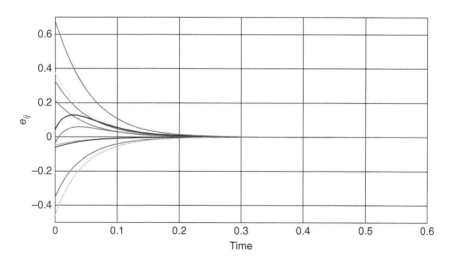

Figure 2.36 Dynamic formation: distance errors $e_{ij}(t)$, $i, j \in V^*$.

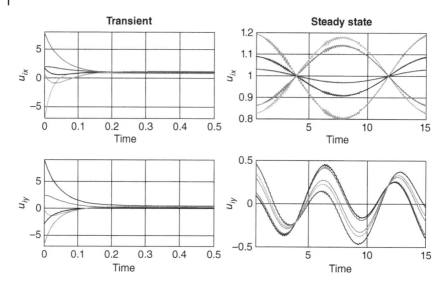

Figure 2.37 Dynamic formation: control inputs $u_i(t)$ $i = 1, \ldots, 5$.

2.7 Notes and References

Several results on formation control of multi-agent systems based on the single-integrator model that consider the stabilization of the inter-agent distance dynamics to desired distances have appeared in the literature. The first to use the gradient descent control law for the 2D formation acquisition problem were [23, 46]. A similar 2D formation acquisition controllers was discussed in [47], whereas 3D formations were considered in [48]. In [49], the authors provided a unified approach for the exponential stability analysis of minimally and nonminimally rigid formations. In [50, 51], 2D formation maneuvering controllers were introduced that only considered swarm translation. 2D formation maneuvering (translation only) and target interception schemes were designed in [2]. In [52], the 2D translational maneuvering scheme involved a leader with a constant velocity command and followers who track the leader while maintaining the formation shape. The control law consisted of the standard gradient descent formation acquisition term plus an integral term to ensure zero steady-state error with respect to the velocity command. In all the above results, the desired formation is assumed static. A related problem is when the control objective is the stabilization to a static formation with a desired orientation. This problem was recently addressed in [53], where the controller included a term to perform orientation stabilization in addition to the gradient descent formation acquisition control. A minimal number of agents should have knowledge of a global coordinate system for this to occur. For other work that addresses this problem, see [53]. Recently, an interesting approach for ensuring convergence to a unique formation (i.e., without ambiguities) was proposed in [54] where the desired formation is defined with both distance and signed area constraints.

To the best of our knowledge, the Lyapunov function candidate (2.10) first appeared in [55], where it was used in the stability analysis of triangular formations. The local nature of our stability results in Euclidean space is inherent to the formation control approach based on rigid graph theory; see, e.g., [23, 46, 47, 56, 57]. Even in the case of three agents attempting to form a triangle, the stability set excludes the situation where the agents are initially collocated or collinear. In fact, the invariance of the collinear set appears to have been first shown in [23]. For n-agent formations, the size of the stability set likely decreases with increasing n since the number of possible framework ambiguities grows. The condition on the initial distance error given in the theorems of this chapter is a sufficient condition for stability, and simply indicates that a stability set exists. In practice, the stability set is best quantified by running simulations for different initial conditions since its size is likely larger than that provided by the sufficient condition.

The directionality of the information exchange among agents is an important design factor. This issue is of practical importance since it relates to the number of communication, sensing, and/or control channels of the multi-agent system. In the case of bidirectional information exchange, a pair of agents concurrently controls the distance between them, whereas only one agent in the pair is responsible for this task in the unidirectional case. In terms of graph theory, bidirectional (resp., unidirectional) formation controllers are based on undirected (resp., directed) graphs. Undirected formation controllers have built-in redundancy, providing robustness. However, it can also lead to instability in the formation acquisition if agent pairs use slightly different values for the distance between them due to measurement errors [31]. In [58, 59] it was shown that this measurement mismatch causes a distortion of the formation from its desired shape and a circular (resp., helical) orbit of the distorted formation in 2D (resp., 3D). One possible remedy for this problem is to have the agents communicate their respective measurements to one another and then use a common value for control (e.g., the average of the two measurements). A more rigorous solution was proposed in [60] by using an estimator-based gradient control law to make the system robust against measurement inconsistency. Yet another solution is to use a directed graph-based controller since it reduces the overall number of communication/sensing/control channels while avoiding the potential conflict between a pair of agents trying to achieve the same objective. However, in directed graphs it is possible to have cycles in the pathways, which are more challenging to control and can lead to formation instability [57]. Therefore, the issue of cyclic versus acyclic graphs is an important consideration for directed formation control.

While the results in this book use undirected graphs, the directed graph problem has also received considerable attention. The idea of applying input-to-state stability to study the stability properties of directed acyclic formation graphs was discussed in [61], but without considering the control design problem. For directed graphs, the traditional notion of rigidity is not enough to maintain the formation structure. In [31], the concept of rigidity of a directed graph was first discussed and a formation control law was proposed that assumes the global position of the leader and first-follower agents are known in a 2D acyclic graph. A method for analyzing the rigidity of 2D directed formations that have a leader–follower structure was provided in [62] without consideration of the control design problem. The concept of rigidity of a directed graph was refined and formalized in [63, 64] by introducing the notions of *constraint consistence* and *persistence*. The intuitive meaning of a constraint consistent graph is that every agent is able to satisfy all its distance constraints when all the others are trying to do the same. A directed graph that is both rigid and constraint consistent is called persistent The notion of persistence was introduced in [63] for 2D graphs and generalized to higher dimensional graphs in [64]. The problem of a triangular formation with cyclic ordering was considered in [65] and a locally exponentially stabilizing control law was proposed based on measurements of all inter-agent distances but only one inter-agent angle. A generalized control law for cyclic triangular formations was presented in [66]. In [7], a control law was designed for directed planar formations of n agents modeled by a minimally persistent graph with a leader-first-follower topology. A controller for n-agent persistent formations whose graph is constructed by a sequence of Henneberg insertions was proposed in [23], and the local asymptotic stability of the desired formation was proven using the linearized dynamics. The stability of the control from [23] was analyzed in [55] using differential geometry for a three-agent cyclic formation. In [57], the result from [7] was generalized to also include minimally persistent graphs with leader-remote-follower and coleader topologies, which are inherently cyclic. In [67], a discontinuous controller was proposed to achieve finite-time convergence to the desired formation for acyclic persistent graphs.

While the above discussion was focused on formation control approaches based on rigid graph theory and controlling inter-agent distances, other approaches can be found in the literature. The treatment of multi-agent formations as a consensus control problem was discussed in [39] where the information state, whose value is to be negotiated by the consensus algorithm, is the center of the formation. A broad literature review of different formation control schemes is presented in Section 2.3 of [4]. In [2], formation acquisition controllers were designed using the gradient of a general potential function that represents the repulsion/attraction between agents and is chosen based on the desired inter-agent distances. Formation maneuvering and target interception schemes were also designed in [2] by adding appropriate terms to the gradient law. Formation acquisition (static and dynamic) and formation maneuvering protocols were introduced in [68] that ensure convergence to the desired formation in finite time. A complex graph Laplacian scheme is presented in [69] for the formation acquisition problem where the formation scale is determined by two leader agents. The idea of triangular formation acquisition using only inter-agent angle measurements was proposed in [70]. In [56], a formation control strategy was demonstrated that is based on the distributed estimation of the agent absolute positions from relative position measurements when the inter-agent sensing/communication graph has a spanning tree.

Relatively few results exist for the general formation maneuvering (translation and rotation) problem. A 3D formation maneuvering control law was presented in [71] but without considering the stability of the formation acquisition. In [72] a controller was proposed that can steer the entire formation in rotation and/or translation in 3D. The rotation component was specified relative to a body-fixed frame whose origin is at the centroid of the desired formation and needs to be known.

The material in this chapter is mostly based on the work in [73–75]. The observers used in the flocking problem of Section 2.3 are based on [76, 77].

3

Double-Integrator Model

In this chapter, we re-discuss the class of formation controllers presented in Chapter 2 in the context of a slightly more refined model, viz., the *double-integrator model*. We will follow the same format as the previous chapter for ease of correlation.

The double-integrator model accounts for the agent *acceleration* by treating the agent as a point mass. Therefore, it can be considered a very simple *dynamic* model for omnidirectional robots. Given a system of n agents, the equations of motion for the double-integrator model are

$$\dot{q}_i = v_i \tag{3.1a}$$

$$\dot{v}_i = u_i, \quad i = 1, \dots, n \tag{3.1b}$$

where $v_i \in \mathbb{R}^m$ represents the velocity of the ith agent with respect to an Earth-fixed coordinate frame, $u_i \in \mathbb{R}^m$ is the acceleration-level control input, and q_i is defined as in (2.1). Since the agent velocity is now a system *state* rather than the control input, the formation control laws in this chapter will be a function of the agent velocities in addition to the positions.

Note that the system transfer function matrix is now $G_i(s) = 1/s^2 I_m$, which gives rise to the model name. Since the only difference between this transfer function and (2.2) is an additional integrator, the extension of the single-integrator-based control laws to (3.1) is rather seamless if one exploits the integrator backstepping methodology (see Appendix C.6).

As in Section 2.1, we begin by deriving the distance error dynamics. To this end, we use (2.6) and (3.1a) to obtain

$$\dot{e}_{ij} = \frac{\tilde{q}_{ij}^T(v_i - v_j)}{e_{ij} + d_{ij}}. \tag{3.2}$$

Differentiating (2.10) along (3.2) gives

$$\dot{W} = \frac{1}{2} z^T \dot{z} = z^T R(\tilde{q}) v \tag{3.3}$$

where $v = [v_1, \dots, v_n] \in \mathbb{R}^{mn}$.

Formation Control of Multi-Agent Systems: A Graph Rigidity Approach, First Edition.
Marcio de Queiroz, Xiaoyu Cai, and Matthew Feemster.
© 2019 John Wiley & Sons Ltd. Published 2019 by John Wiley & Sons Ltd.
Companion website: www.wiley.com/go/dequeiroz/formation_control

Given that v in (3.3) cannot be directly prescribed since it is a system state, we follow the backstepping technique and introduce the following variable

$$s = v - v_f \tag{3.4}$$

where $v_f \in \mathbb{R}^{mn}$ denotes the *fictitious* (or desired) velocity input, which will be specified later. The variable s quantifies the error between the actual agent velocity and the desired velocity-level input. The design of v_f will be *problem-specific*, and will come from the velocity-level control laws of Chapter 2. That is, generally speaking, $v_f = u^{SI}$ where the superscript SI stands for one of the control input designs for the single-integrator model. The block diagrams in Figure 3.1 illustrate the relationship between the control designs for the single- and double-integrator models. As one can see, the velocity-level, position control algorithms from Chapter 2 will be embedded in the acceleration-level, velocity control loop to be designed in this chapter.

Due to the new error variable (3.4), we introduce the augmented Lyapunov function candidate

$$W_d(e, s) = W(e) + \frac{1}{2}s^\mathsf{T}s \tag{3.5}$$

where W was defined in (2.10). Notice that W is a potential energy-like term since it is only position dependent, whereas $\frac{1}{2}s^\mathsf{T}s$ is a kinetic energy-like term due to its dependence on velocity. Therefore, W_d captures the total energy of the double-integrator model formation.

After taking the time derivative of (3.5), we obtain

$$\begin{aligned}
\dot{W}_d &= z^\mathsf{T}R(\tilde{q})v + s^\mathsf{T}\dot{s} \\
&= z^\mathsf{T}R(\tilde{q})(s + v_f) + s^\mathsf{T}(u - \dot{v}_f) \\
&= z^\mathsf{T}R(\tilde{q})v_f + s^\mathsf{T}(u + R^\mathsf{T}(\tilde{q})z - \dot{v}_f)
\end{aligned} \tag{3.6}$$

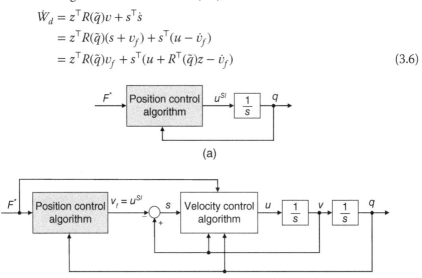

Figure 3.1 Relationship between the (a) single- and (b) double-integrator control designs.

where (3.3), (3.1b), and (3.4) were used. Equation (3.6) is the analogue of (2.12) since it will be the starting point for all double-integrator control designs as (2.12) was for the single-integrator designs.

3.1 Cross-Edge Energy

Before presenting the formation controllers, we need to discuss a complication in the stability analysis of the closed-loop system that arises from the double-integrator model. Specifically, this complication is related to the avoidance of flip ambiguities.

Recall that for the single-integrator model, the position of the initial formation needs to be restricted to prevent convergence to a flip ambiguity since the velocity-level control input is designed to promote convergence to $\mathrm{Iso}(F^*)$ or $\mathrm{Amb}(F^*)$, whichever is closer at $t = 0$. Unfortunately, this condition is not sufficient for the double-integrator model. In this case, the agents' *velocity* will also affect the convergence since it is a system state. This idea is conceptually illustrated by Figure 3.2. Note that even if the formation position is closer to $\mathrm{Iso}(F^*)$, the formation will overcome the energy barrier and converge to $\mathrm{Amb}(F^*)$ if its velocity is large enough. In other words, the total formation energy is now affected by the combination of potential energy and kinetic energy. The implication of this for stability is that a restriction also needs to be imposed on the initial velocity of the formation, which means that we need to limit the initial *total energy* of the formation.

While the need for an upper bound on the initial energy of the formation is evident, its precise value is difficult to calculate in general. For simple formations, one may be able to calculate a conservative value for the energy upper bound as illustrated next. Consider the desired triangular formation in Figure 3.3 along with one of its flipped versions. Note that a flip may occur whenever an agent has enough energy to cross the edge connecting the two other agents, e.g., agent 1 crossing edge (2, 3). Once the agent crosses the edge, it is closer to $\mathrm{Amb}(F^*)$ and may be attracted to this undesired equilibrium.

Figure 3.2 Energy landscape where the formation is at position q with velocity v.

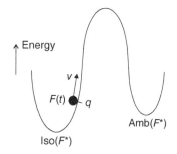

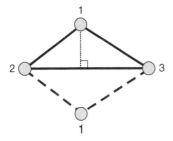

Figure 3.3 Desired formation (solid line) and a flip ambiguity (dashed line).

The question is then: What is the minimum energy needed for this to happen? Hereafter, we refer to this minimum energy as the *cross-edge energy*, E_c.

A conservative estimate for the cross-edge energy can be made by using the following observations: (i) the cross-edge energy is related to the energy that drives the agents to a collinear formation and (ii) the minimum collinearity energy is given by the agent with the smallest distance to its cross-edge, e.g., the dotted line in Figure 3.3. These rules facilitate the cross-edge energy estimation because they are only *position dependent*. Furthermore, we have from (3.5) and (2.10) that $W_d \geq W = \frac{1}{4}z^\top z$, which is also only position dependent. That is, a sufficient condition for E_c can be determined by calculating the minimum value of W when the three agents are collinear. For example, let $d_{12} = d_{13} = \sqrt{2}$ and $d_{23} = 2$. When agent 1 is collinear with agents 2 and 3, we have that $\|\tilde{q}_{12}\| + \|\tilde{q}_{13}\| = \|\tilde{q}_{23}\|$. For notational convenience, we use $q \in C$ where $q = [q_1, q_2, q_3]$ to denote that the agents are collinear. Therefore,

$$E_c = \min_{q \in C} W = \min_{q \in C} \frac{1}{4}(z_{12}^2 + z_{13}^2 + z_{23}^2)$$

$$= \min_{q \in C} \frac{1}{4}\left[(\|\tilde{q}_{12}\|^2 - d_{12}^2)^2 + (\|\tilde{q}_{13}\|^2 - d_{13}^2)^2 + (\|\tilde{q}_{23}\|^2 - d_{23}^2)^2\right]$$

$$= \min \frac{1}{4}\left[(\|\tilde{q}_{12}\|^2 - 2)^2 + ((\|\tilde{q}_{23}\| - \|\tilde{q}_{12}\|)^2 - 2)^2 + (\|\tilde{q}_{23}\|^2 - 4)^2\right].$$

It can be found that the above function reaches a minimum at $\|\tilde{q}_{12}\| = \|\tilde{q}_{23}\|/2 = \sqrt{10}/3$ and $E_c = 0.444$. This means that if $W_d(0) \leq E_c$, the agents will not converge to the flip ambiguity.

Notice that the condition $W_d(0) \leq E_c$ imposes a trade-off between the initial distance error and the initial velocity error. The larger the initial distance error, the smaller the initial velocity error needs to be, and vice versa. Based on (3.4), a small s implies that the agents' velocities are close to v_f, which is the desired velocity that ensures convergence to $\mathrm{Iso}(F^*)$.

For formations with $n > 3$, one may apply the above estimation method by triangulating the framework and comparing the cross-edge energy of each triangle to estimate E_c. For example, consider the infinitesimally rigid framework in Figure 3.4. The agents most likely to flip are agents 2 and 6 about cross-edges $(1, 3)$ and $(1, 5)$, respectively, since they only have two edges

Figure 3.4 Triangulated hexagon framework.

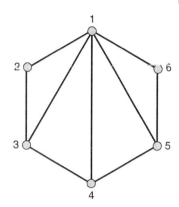

(constraints) each. Thus, $E_c = \min\{E_{c2}, E_{c6}\}$ where E_{ci} denotes the cross-edge energy of agent i. Note that *higher order* flips are also possible, but they would require more energy than aforementioned single-agent flips. For example, agents $\{2, 3\}$ or $\{5, 6\}$ could simultaneously also flip about cross-edge $(1, 4)$, or agents $\{2, 3, 4, 5, 6\}$ could simultaneously flip about agent 1, leading to a full reflection of the formation.

3.2 Formation Acquisition

The formation acquisition controller for (3.1) will have the general form $u_i = u_i(q_i - q_j, v_i - v_j, v_i, d_{ij})$, $i = 1, \ldots, n$ and $j \in \mathcal{N}_i(E^*)$ where $\mathcal{N}_i(\cdot)$ was defined in (1.2). Based on (3.6), the following theorem introduces the control law that solves the formation acquisition problem.

Theorem 3.1   Given the formation $F(t) = (G^*, q(t))$, let the initial conditions be such that $(e(0), s(0)) \in \Omega_1 \cap \Omega_2 \cap \Omega_3$ where Ω_1 and Ω_2 were defined in (2.14),

$$\Omega_3 = \{e \in \mathbb{R}^l, s \in \mathbb{R}^{mn} \mid W_d \leq E_c\}, \tag{3.7}$$

and E_c is the total cross-edge energy of the formation. Then, the control

$$u = -k_a s + \dot{v}_f - R^\top(\tilde{q})z, \tag{3.8}$$

where

$$v_f = u_a, \tag{3.9}$$

u_a was defined in (2.15), and $k_a > 0$ is a user-defined control gain, renders $(e, s) = 0$ exponentially stable and ensures that (1.26) is satisfied.

Proof: Substituting (3.8) and (3.9) into (3.6) yields

$$\dot{W}_d = -k_v z^\top R(\tilde{q}) R^\top(\tilde{q}) z - k_a s^\top s. \tag{3.10}$$

Following the arguments used in the proof of Theorem 2.1, we have that

$$\dot{W}_d \le -k_v \lambda_{\min}(RR^\top)z^\top z - k_a s^\top s \le -\min\{2k_a, 4k_v \lambda_{\min}(RR^\top)\} W_d \quad (3.11)$$

for $e(0) \in \Omega_1$. From (3.5) and (3.11), we know that $(e, s) = 0$ is exponentially stable for $e(0) \in \Omega_1$ from Corollary C.1, and therefore, $F(t) \to \mathrm{Iso}(F^*)$ or $F(t) \to \mathrm{Amb}(F^*)$ as $t \to \infty$ for $e(0) \in \Omega_1$. Now, if the initial conditions are chosen such that $(e(0), s(0)) \in \Omega_1 \cap \Omega_2 \cap \Omega_3$, we know that the formation starts closer to $\mathrm{Iso}(F^*)$ than $\mathrm{Amb}(F^*)$ and $W_d(e(0), s(0)) < E_c$. Since $\dot{W}_d \le 0$, we know that $W_d(e(t), s(t)) < E_c$ for all $t > 0$, indicating that the formation energy is always less than the minimum energy required for a flip to occur. Thus, we have that (1.26) holds for $(e(0), s(0)) \in \Omega_1 \cap \Omega_2 \cap \Omega_3$. $\qquad\square$

The expression for $\dot{v}_f$ in (3.8) is given by

$$\dot{v}_f = -k_v \dot{R}^\top z - k_v R^\top \dot{z} \quad (3.12)$$

where from (1.15)

$$\dot{R}(\tilde{q}) = R(\tilde{v}), \quad (3.13)$$

$\tilde{v} = [\dots, v_i - v_j, \dots] \in \mathbb{R}^l$, $(i, j) \in E^*$, and from (3.3)

$$\dot{z} = 2R(\tilde{q})v. \quad (3.14)$$

The control (3.8)–(3.9) can be written element-wise as

$$u_i = -k_a v_i - \sum_{j \in \mathcal{N}_i(E^*)} [(k_a k_v + 1)\tilde{q}_{ij} z_{ij} + k_v(z_{ij} I_2 + 2\tilde{q}_{ij}\tilde{q}_{ij}^T)\tilde{v}_{ij}] \quad (3.15)$$

for $i = 1, \dots n$ and

$$\tilde{v}_{ij} = v_i - v_j, \quad (i, j) \in E^*. \quad (3.16)$$

This control is decentralized in the sense of Definition 1.1 since its implementation only requires each agent to measure its own velocity and the relative position and relative velocity to neighboring agents. The agent's velocity can be measured using onboard sensors such as an odometer and a compass.

3.3   Formation Maneuvering

The formation maneuvering control law for the double-integrator model (3.1a)–(3.1b) is simply a combination of the designs in Sections 2.2 and 3.2. Specifically, u is given by (3.8) with

$$v_f = u_a + v_d \quad (3.17)$$

where the formation maneuvering velocity v_d was specified in (2.24). Note that (3.17) is exactly the right-hand side of (2.23).

We will not present the formal statement and proof of this result, but only discuss the aspects in which it differs from the proofs of Theorems 2.2 and 3.1. This is namely the proof that (1.28) holds. First, after substituting (3.17) into (3.6), the proofs of the exponentially stability of $(e, s) = 0$ and (1.26) are straightforward given that $R(\tilde{q})v_d = 0$ (see (1.20) and (2.24)). Now, since $e(t) \to 0$ as $t \to \infty$, we know from (2.9) that $z(t) \to 0$ as $t \to \infty$. Since $R(\tilde{q})$ is bounded, then $u_a(t) \to 0$ as $t \to \infty$ from (2.15). Therefore, we have that $v_f(t) \to v_d(t)$ as $t \to \infty$ from (3.17). Since we know $s(t) \to 0$ as $t \to \infty$, it follows from (3.4) that $v(t) - v_f(t) \to 0$ as $t \to \infty$. Therefore, $v_i(t) - v_{di}(t) \to 0$ as $t \to \infty$, $i = 1, ..., n$, which is the same as (1.28) due to (3.1a).

The term $\dot{v}_f$ in (3.8) will contain additional terms from the derivative of v_d. Specifically, from (2.24), we have that

$$\dot{v}_{di} = \dot{v}_t + \dot{\omega} \times \tilde{q}_{in} + \omega \times \tilde{v}_{in}, \quad i = 1, \dots n \tag{3.18}$$

where $\dot{v}_t \in \mathbb{R}^3$ denotes the desired translational acceleration and $\dot{\omega} \in \mathbb{R}^3$ is the desired angular acceleration for the virtual rigid body. Therefore, for the double-integrator model, v_t and ω need to be continuously differentiable functions of time with bounded first derivative for the control input to be continuous and bounded. Note that element-wise the formation maneuvering control law is simply made up of the sum of the right-hand sides of (3.15) and (3.18). Like v_t and ω, the signals $\dot{v}_t$ and $\dot{\omega}$ can be stored on each agent's onboard computer since they are typically known a priori.

3.4 Target Interception with Unknown Target Acceleration

Solving the target interception problem for the double-integrator model requires a more elaborate solution than the one presented in Section 2.4 for the single-integrator model. Here, we consider that the target position $q_T(t)$ is twice continuously differentiable and $q_T(t), \dot{q}_T(t), \ddot{q}_T(t) \in \mathcal{L}_\infty$. We also assume the signals $q_T - q_n$, $\dot{q}_T - \dot{q}_n$, $\dot{q}_n$, and $\dot{q}_T$ are known and can be broadcast from the leader to the followers; however, the signal $\ddot{q}_T$ is *unknown*. A variable structure-type control term will be used to compensate for the unknown target acceleration. As a result, the right-hand side of the resulting error system dynamics will be discontinuous, requiring us to apply some ideas from Lyapunov stability of nonsmooth systems. As in Section 2.4, we let $v_T := \dot{q}_T$ to simplify the notation.

Theorem 3.2 Let the initial conditions be such that $(e(0), s(0)) \in \Omega_1 \cap \Omega_2 \cap \Omega_3$ where Ω_1 and Ω_2 were defined in (2.14) and Ω_3 was defined in (3.7). Consider the control

$$u = -k_a s + \dot{u}_a + \mathbf{1}_n \otimes k_T(v_T - v_n) - k_s \, sgn(s) - R^\mathsf{T}(\tilde{q})z \tag{3.19}$$

where s was defined in (3.4),

$$v_f = u_a + \mathbf{1}_n \otimes h, \tag{3.20}$$

$$h = k_T e_T + v_T, \tag{3.21}$$

u_a was defined in (2.15), e_T was defined in (2.44), $k_s \geq \sqrt{n}\|\dot{v}_T\|_{\mathcal{L}_\infty}$, and $k_T > 0$. Then, (3.19) renders $(e, s) = 0$ asymptotically stable and ensures that (1.26) and (1.29) are satisfied.

Proof: First, notice that the differential equations describing the (e, s)-error system dynamics in a closed loop with (3.19)–(3.21) have a discontinuous right-hand side due to the term $\text{sgn}(s)$ in (3.19). That is, if $\dot{\xi} = f(\xi, t)$ denotes the closed-loop system where $\xi = [e, s]$, then $f(\xi, t)$ is continuous everywhere except in the set $\{(\xi, t) \mid s = 0\}$. For such a system, a generalized solution exists by embedding the differential equations into the differential inclusions $\dot{\xi} \in K[f](\xi, t)$. In this case, the time derivative of (3.5) is given by [42]

$$\dot{W}_d \stackrel{a.e.}{\in} \frac{\partial W_d}{\partial \xi} K[f](\xi, t)$$
$$\subset z^\top R(\tilde{q}) v_f + s^\top (u + R^\top(\tilde{q})z - \dot{v}_f) \tag{3.22}$$

where (3.6) was used. Substituting (2.15), (3.19), (3.20), and (3.21) into (3.22) and then applying Property 1.1 gives [42]

$$\dot{W}_d \subset -k_v z^\top R R^\top z - k_a s^\top s - s^\top (k_s \, \text{sgn}(s) + \mathbf{1}_n \otimes \dot{v}_T)$$
$$= -k_v z^\top R R^\top z - k_a s^\top s - s^\top (k_s \, \text{SGN}(s) + \mathbf{1}_n \otimes \dot{v}_T)$$
$$\leq -k_v z^\top R R^\top z - k_a s^\top s + \|s\|(\sqrt{n}\|\dot{v}_T\|_{\mathcal{L}_\infty} - k_s) \tag{3.23}$$

where $\text{sgn}(\cdot)$ and $\text{SGN}(x)$ were defined in (2.37) and (2.42), respectively.

For $k_s \geq \sqrt{n}\|\dot{v}_T\|_{\mathcal{L}_\infty}$, (3.23) reduces to (3.11) so $\dot{W}_d$ is negative definite for $e(0) \in \Omega_1$. Therefore, from Theorem C.6, we know that $(z, s) = 0$ is asymptotically stable. Since W_d is positive definite in e, we know that $(e, s) = 0$ is asymptotically stable for $e(0) \in \Omega_1$. The proof of (1.26) for $(e(0), s(0)) \in \Omega_1 \cap \Omega_2 \cap \Omega_3$ now follows from the same arguments used in the proof of Theorem 3.1.

Next, from (3.20), we have

$$v_{fn} = u_{an} + v_T + k_T e_T \tag{3.24}$$

where the subscript n denotes the nth element of the corresponding vector. Differentiating (2.44) and applying (3.24) yields

$$\dot{e}_T = v_T - v_n = v_T - (v_{fn} + s_n)$$
$$= -k_1 e_T + r \tag{3.25}$$

where $r := -s_n - u_{an}$. Since (3.25) is a *stable* linear system with input r and output e_T, the output will converge to zero if the input converges to zero (see Theorem C.1). Given that $(z, s) = 0$ is asymptotically stable, we know that $(s(t), z(t)) \to 0$ as $t \to \infty$ and therefore $u_a(t) \to 0$ as $t \to \infty$. As a result, $r(t) \to 0$ as $t \to \infty$ and, from (3.25), $e_T(t) \to 0$ as $t \to \infty$. Finally, since (1.26) implies that $q_n(t) \in \text{conv}\{q_1(t), q_2(t), ..., q_{n-1}(t)\}$ as $t \to \infty$ due to the manner in which F^* is constructed for the target interception problem, we conclude from the convergence of e_T to zero that (1.29) holds. $\qquad\square$

A few observations are in order concerning the structure of (3.19)–(3.21). First, $\dot{v}_f$ is not included in (3.19) as it is in (3.8) because the derivative of (3.20) is a function of the unknown signal $\dot{v}_T$. Hence, only the measurable terms of $\dot{v}_f$ appear in (3.19). Since $\dot{v}_T$ cannot be directly cancelled by the control, it is instead dominated by the variable structure term $k_s\text{sgn}(s)$ as shown in (3.23). Second, comparing (2.54) and (3.21), notice the absence of the term $-u_{an}$ in the latter. Unlike the control in Theorem 2.4, the presence of this term in (3.21) is not necessary for proving the converge of e_T to zero. If $-u_{an}$ was included (3.21), the above stability analysis would still hold with the exception that the auxiliary variable r in (3.25) would become simply $r = -s_n$.

When expressed element-wise, the control (3.19)–(3.21) takes the form

$$
u_i = - k_a v_i + k_a(v_T + k_T e_T) + k_T(v_T - v_n)
$$
$$
- \sum_{j \in \mathcal{N}_i(E^*)} [k_v(z_{ij}I_2 + 2\tilde{q}_{ij}\tilde{q}_{ij}^T)\tilde{v}_{ij} + (k_a k_v + 1)\tilde{q}_{ij}z_{ij}]
$$
$$
- k_s \, \text{sgn}(v_i - v_T - k_T e_T + k_v \sum_{j \in \mathcal{N}_i} \tilde{q}_{ij}z_{ij}).
$$

As one can see, the ith agent's control input is dependent on its own velocity and the relative position/velocity to neighboring agents, e_T, v_T, and v_n.

3.5 Dynamic Formation Acquisition

When solving the dynamic formation acquisition problem (see Problem 4 in Section 2.5) for the double-integrator model, we require that the time-varying distance $d_{ij}(t)$ be twice continuously differentiable and $d_{ij}(t), \dot{d}_{ij}(t), \ddot{d}_{ij}(t) \in \mathcal{L}_\infty$ for the control law to be continuous and bounded.

Similar to the formation maneuvering control law of this chapter, the dynamic formation acquisition control input will take the form of (3.8) but with the problem-specific design for v_f. That is, v_f is set to the right-hand side of (2.71) for dynamic formation acquisition.

The term $\dot{v}_f$ in (3.8) can be explicitly calculated from (2.71) as follows

$$
\dot{v}_f = \dot{R}^+(\bar{q})(-k_v z + d_v) + R^+(\bar{q})(-k_v \dot{z} + \dot{d}_v) \tag{3.26}
$$

where d_v was defined in (2.70),

$$\dot{d}_v = [..., \dot{d}_{ij}^2 + d_{ij}\ddot{d}_{ij}, ...], (i,j) \in E^*,$$

$$\dot{z} = 2(R(\tilde{q})v - d_v),$$

$$\dot{R}^+ = \dot{R}^\mathsf{T}(RR^\mathsf{T})^{-1} - R^\mathsf{T}(RR^\mathsf{T})^{-1}\frac{d(RR^\mathsf{T})}{dt}(RR^\mathsf{T})^{-1},$$

and $\dot{R}$ was defined in (3.13). It is not difficult to see that (3.26) is a function of $\tilde{q}_{ij}$, $\tilde{v}_{ij}$, d_{ij}, $\dot{d}_{ij}$, and $\ddot{d}_{ij}$ for $(i,j) \in E^*$. This control also suffers from the coupling issue discussed in Section 2.5 due to the presence of the pseudoinverse matrix R^+ in (2.71) and (3.26).

The proof of stability uses the same Lyapunov function candidate (3.5) and combines the arguments from the proofs of Theorems 2.5 and 3.1. A sketch of the proof is as follows. Substituting (3.8) and (2.71) into (3.6) yields

$$\dot{W}_d = -k_v z^\mathsf{T} z - k_a s^\mathsf{T} s \leq -2\min(k_v, k_a)W_d \tag{3.27}$$

for $e(t) \in \Omega_1$ from which we conclude that $(e,s) = 0$ is exponentially stable for $e(0) \in \Omega_1$ in the same vein of Theorem 3.1. The proof of (2.66) for $(e(0), s(0)) \in \Omega_1 \cap \Omega_2 \cap \Omega_3$ proceeds as in Theorem 3.1.

As in the single-integrator case, formation maneuvering can be performed concurrently with dynamic formation acquisition by setting v_f to the right-hand side of (2.73). The derivative of v_f will then be given by (3.26) plus $\dot{v}_d$ as defined in (3.18).

3.6 Simulation Results

The MATLAB simulations in this chapter will show the agents performing formations in 3D based on the model in (3.1).

3.6.1 Formation Acquisition

An eight-agent simulation was conducted to demonstrate the performance of control law (3.8). The desired formation F^* was the cube with edge length of 2 shown in Figure 3.5 where $q_1^* = [-1, -1, -1]$, $q_2^* = [1, -1, -1]$, and so on. The desired framework was made minimally rigid and infinitesimally rigid by introducing 18 ($= 3 \times 8 - 6$) edges with edge set

$$E^* = \{(1,2), (1,3), (1,4), (1,5), (1,8), (2,3), (2,5), (2,6), (3,4),$$

$$(3,6), (3,7), (4,7), (4,8), (5,6), (5,8), (6,7), (6,8), (7,8)\}.$$

That is, each face of the cube was "triangulated" by adding a diagonal edge. The desired distances for $(i,j) \in E^*$ were given by $d_{12} = d_{14} = d_{15} = d_{23} = d_{26} =$

Figure 3.5 Formation acquisition: desired formation F^*.

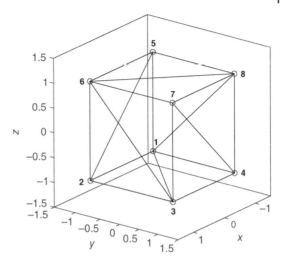

$d_{34} = d_{37} = d_{48} = d_{56} = d_{58} = d_{67} = d_{78} = 2$ and $d_{13} = d_{18} = d_{25} = d_{36} = d_{47} = d_{68} = 2\sqrt{2}$.

The initial conditions of the agents were randomly selected by

$$q_i(0) = q_i^* + \text{rand}(0, 1) - 0.51_3 \quad \text{and}$$
$$v_i(0) = 2(\text{rand}(0, 1) - 0.51_3), \quad i = 1, \dots, 8 \tag{3.28}$$

where the function $\text{rand}(0, 1)$ generates a random 3×1 vector whose elements are uniformly distributed on the interval $(0, 1)$. Both control gains k_v and k_a were set to 1.

The agent trajectories in space as they converge to the desired cube formation are shown in Figure 3.6. All 28 inter-agent distance errors $e_{ij}(t), i, j \in V^*$ are depicted in Figure 3.7, confirming the acquisition of the desired formation. In Figure 3.8 we plot the x-, y-, and z-direction components of the velocity-related error variable s defined in (3.4), which according to Theorem 3.1 should converge to zero. Finally, Figure 3.9 shows the components of the acceleration-level control inputs.

3.6.2 Dynamic Formation Acquisition with Maneuvering

This simulation combines formation maneuvering with dynamic formation acquisition as discussed at the end of Section 3.5. For this case, the desired dynamic formation $F^*(t)$ was set to a cube that expanded and contracted uniformly over time. The desired formation was initialized as the cube with edge length of 2 shown in Figure 3.5. The eight vertices of the cube represented the followers while agent 9 at the geometric center of the cube was the leader through which the rotation axis passed. We ensured the desired framework

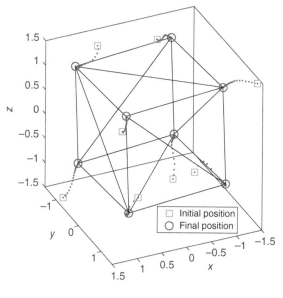

Figure 3.6 Formation acquisition: agent trajectories $q_i(t)$, $i = 1, ..., 8$.

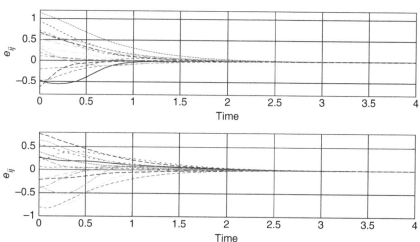

Figure 3.7 Formation acquisition: distance errors $e_{ij}(t)$, $i, j \in V^*$.

was minimally rigid and infinitesimally rigid by imposing $21 \ (= 3 \times 9 - 6)$ edges with all followers connected to the leader. The edge set was selected as

$$E^* = \{(1,4),(1,5),(1,9),(2,3),(2,6),(2,9),(3,4),(3,7),(3,9),(4,5),(4,8),$$
$$(4,9),(5,6),(5,8),(5,9),(6,7),(6,8),(6,9),(7,8),(7,9),(8,9)\}.$$

$$(3.29)$$

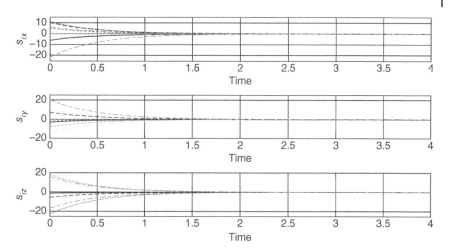

Figure 3.8 Formation acquisition: velocity errors $s_i(t)$, $i = 1, \ldots, 8$.

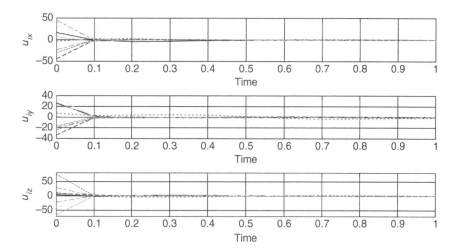

Figure 3.9 Formation acquisition: control inputs $u_i(t)$, $i = 1, \ldots, 8$.

The desired formation was made dynamic by setting the vertex coordinates to $q_i^*(t) = q_i^*(0)\rho(t)$ where ρ was defined in (2.81). The maneuvering velocity v_d in (2.24) was set to have the following translational and rotational components

$$v_t = [1, \cos t, 0] \quad \text{and} \quad \omega = [1, 1, 1],$$

which results in a screw-like motion for the formation. Control gains k_v and k_a were again set to 1, while initial conditions were chosen according to (3.28).

Snapshots in time of the actual formation are shown in Figure 3.10, where the dotted line marks the trajectory of the leader. The 36 inter-agent distance errors

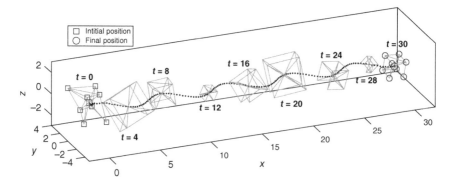

Figure 3.10 Dynamic formation acquisition with maneuvering: snapshots of $F(t)$ at different instants of time.

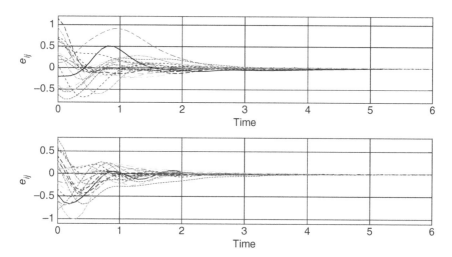

Figure 3.11 Dynamic formation acquisition with maneuvering: distance errors $e_{ij}(t)$, $i, j \in V^*$.

are given in Figure 3.11, while the velocity errors are shown in Figure 3.12. In Figure 3.13, the control inputs are depicted.

3.6.3 Target Interception

In this simulation, the desired formation F^* was set to the framework in Figure 3.5 with edge set (3.29). The leader, who is responsible for tracking the target, was agent 9. The velocity of the moving target was chosen as

$$v_T = [0, 1, \cos t]$$

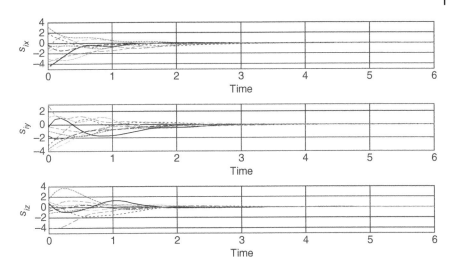

Figure 3.12 Dynamic formation acquisition with maneuvering: velocity errors $s_i(t)$, $i = 1, \ldots, 9$.

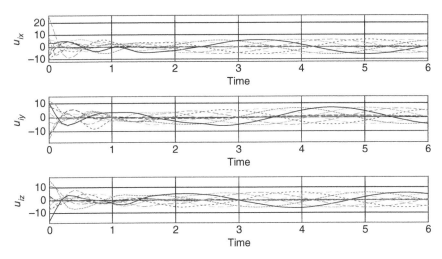

Figure 3.13 Dynamic formation acquisition with maneuvering: control inputs $u_i(t)$, $i = 1, \ldots, 9$.

with initial position $q_T(0) = [0, 2, 0]$. The initial conditions of the agents were chosen according to (3.28), while the control gains in (3.19)–(3.21) were set to $k_s = 3$, $k_a = 0.1$, $k_T = 0.5$, and $k_v = 0.7$.

Figure 3.14 shows the leader intercepting the target while the followers simultaneously surround it with the cube formation. All 36 inter-agent

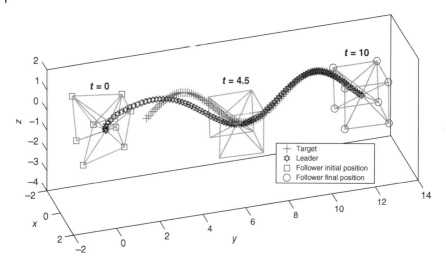

Figure 3.14 Target interception: snapshots of $F(t)$ at different instants of time along with target motion.

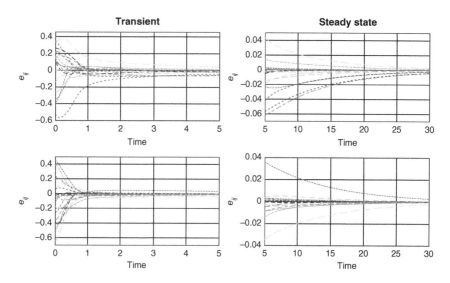

Figure 3.15 Target interception: distance errors $e_{ij}(t)$, $i, j \in V^*$.

distance errors are given in Figure 3.15 with the left (resp., right) plots showing the transient (resp., steady-state) behavior. The directional components of the velocity errors and control inputs of each agent are shown in Figures 3.16 and 3.17, respectively. Note that the chattering-like appearance of the control inputs stems from the discontinuous term sgn(s) in (3.19).

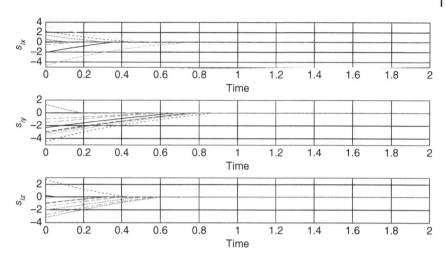

Figure 3.16 Target interception: velocity errors $s_i(t)$, $i = 1, \ldots, 9$.

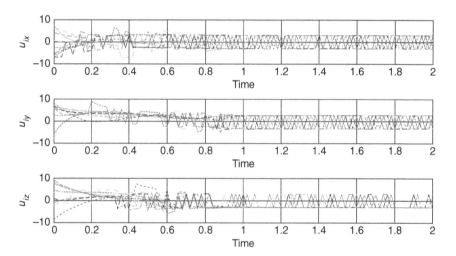

Figure 3.17 Target interception: control inputs $u_i(t)$, $i = 1, \ldots, 9$.

3.7 Notes and References

Formation controllers based on the double-integrator model are not as prevalent as single-integrator-based ones, especially within the realm of inter-agent distance control. As in Chapter 2, the discussion below is mostly focused on results that explicitly or implicitly address the formation control problem.

In [34, 48], the double-integrator, inter-agent distance dynamics for formation acquisition were described as a Hamiltonian system and the local asymptotic stability of undirected formations was achieved under a gradient-like control law. In [46], the gradient formation acquisition law was extended to the double-integrator model for tree formation graphs. Formation acquisition and flocking control systems were studied in [78] where the invariant properties between single- and double-integrator formation systems were established by employing a parameterized Hamiltonian system.

A dynamic formation maneuvering control law was presented in [79] that decouples formation acquisition from maneuvering using virtual bodies and artificial repel-attract potentials. Decoupling was achieved by parameterizing the virtual body motion by the scalar variable whose speed and direction can be prescribed. Recall that the general idea of decoupling formation acquisition and maneuvering was also used in the control designs of Sections 2.2 and 3.3 by exploiting the structure of the rigidity matrix (see (1.20)).

In [80], the time-varying formation problem, which includes formation maneuvering and dynamic formation, was transformed into a consensus problem with respect to a formation center function. A 2D formation maneuvering controller was proposed in [81] where the group leader, who has inertial frame information, passes the information to other agents through a directed path in the graph. A limitation of this control is that it becomes unbounded if the desired formation maneuvering velocity is zero. A synchronization strategy was applied to the formation maneuvering problem in [82] where agents track their individual desired trajectory while synchronizing their relative motions to maintain the desired formation. A consensus scheme was presented in [83] using both the single- and double-integrator models where the formation translation velocity is constant and known to only two leader agents. In [84], existing gradient controllers were modified to ensure finite time formation acquisition and flocking. A similar problem was addressed in [85] but with asymptotic formation acquisition and velocity consensus.

The consensus-type algorithms for formation control proposed in [39] were extended by the authors to the double-integrator dynamics. Several problems were discussed, including consensus conditions for fixed and switching interaction graphs, bounded control effort, and elimination of agent velocity measurements.

The interesting problem of containment control was studied in [86], where the followers move in the convex hull spanned by multiple leaders while the leaders perform formation maneuvers. Experiments using wheeled mobile robots were provided to validate the containment algorithms.

The first use of backstepping to deal with the double-integrator model appeared in [87]. The goal of this work was to extend the single-integrator result of [65] (formation acquisition for three agents with directed graphs) to the double-integrator case with formation translation. Comprehensive coverage of the integrator backstepping control technique can be found in [9].

The work in [88] considered the problem of dynamic formation acquisition with scaling of the formation size, where only a subset of agents know the desired scaling size but all agents know the desired formation shape. Recently, [89] analyzed the influence of mismatches on the measured distances of neighboring agents on the standard gradient-based rigid formation control for double-integrator agents. It was shown that, like the single-integrator case discussed in [58], these mismatches introduce a distorted final shape and a steady-state motion of the formation.

The material in this chapter is based on the work in [74, 75, 90, 91].

4

Robotic Vehicle Model

In this chapter we extend the graph rigidity-based formation control framework to multi-robotic vehicles. As opposed to the simple linear models of Chapters 2 and 3, the agent model here will include the *nonlinear* kinematics and dynamics of the vehicle. Specifically, we will consider a class of robotic vehicles moving in 2D which includes unicycle robots, marine (surface) vessels, underwater vehicles with constant depth, and aircraft with constant altitude.

In the first part of the chapter we will only account for the nonholonomic kinematics of the vehicles and design a velocity-level control law based on the main result from Chapter 2. In the second part, we will include the holonomic vehicle dynamics in the control design. Since the resulting dynamic model will be a *second-order* nonlinear differential equation, the backstepping methodology will again be utilized to embed the velocity-level inputs from Chapter 2 in the torque/force-level control law. Two controllers will be presented in this second part. First, we consider the case where the dynamics are completely known, leading to the design of a fully model-based formation controller. We then assume the model is subject to parametric uncertainty. In this case, we use adaptive control tools to add parameter adaptation to the control law for the purpose of compensating for the unknown parameters.

The discussions in this chapter are limited to the static formation acquisition problem. Extensions to the other formation problems are left as an exercise for the reader.

4.1 Model Description

Consider a heterogenous system of n robotic vehicles moving autonomously on the plane. Figure 4.1 depicts the ith vehicle, where the reference frame $\{X_0, Y_0\}$ is fixed to the Earth. The moving reference frame $\{X_i, Y_i\}$ is attached to the ith vehicle with the X_i axis aligned with its heading (longitudinal) direction, which is given by angle θ_i and measured counterclockwise from the X_0 axis. Point C_i

Formation Control of Multi-Agent Systems: A Graph Rigidity Approach, First Edition.
Marcio de Queiroz, Xiaoyu Cai, and Matthew Feemster.
© 2019 John Wiley & Sons Ltd. Published 2019 by John Wiley & Sons Ltd.
Companion website: www.wiley.com/go/dequeiroz/formation_control

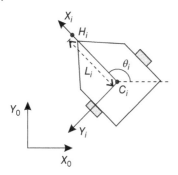

Figure 4.1 The ith robotic vehicle.

denotes the ith vehicle's center of mass, which is assumed to coincide with its center of rotation.

We assume the following model for the vehicles [92, 93]

$$\dot{p}_i = S(\theta_i)\eta_i \tag{4.1a}$$

$$\overline{M}_i\dot{\eta}_i + \overline{D}_i\eta_i = \overline{\tau}_i \tag{4.1b}$$

for $i = 1, \dots, n$, where the first (resp., second) equation describes the vehicle kinematics (resp., dynamics). In (4.1a), $p_i = [x_{ci}, y_{ci}, \theta_i]$ is the position and orientation of $\{X_i, Y_i\}$ relative to $\{X_0, Y_0\}$ (a.k.a. the pose of the robot), $\eta_i = [v_i, \omega_i]$, v_i is the ith robot's translational speed in the direction of θ_i, ω_i is the ith robot's angular speed about the vertical axis passing through C_i, and

$$S(\theta_i) = \begin{bmatrix} \cos\theta_i & 0 \\ \sin\theta_i & 0 \\ 0 & 1 \end{bmatrix}. \tag{4.2}$$

In (4.1b), $\overline{M}_i = \mathrm{diag}(m_i, \overline{I}_i)$, m_i is the ith vehicle mass, $\overline{I}_i$ is the ith vehicle moment of inertia about the vertical axis passing through C_i, $\overline{D}_i \in \mathbb{R}^{2\times 2}$ is the constant damping matrix, and $\overline{\tau}_i \in \mathbb{R}^2$ represents the force/torque-level control input provided by the actuation system.

The main challenge in dealing with (4.1) is that the vehicle kinematics 4.1a is *nonholonomic*[1] since the dimension of the admissible velocity space ($\eta_i \in \mathbb{R}^2$) is smaller than the dimension of the configuration space ($p_i \in \mathbb{R}^3$). This is because the vehicle cannot move in the direction of axis Y_i (e.g., the wheels of a robotic car cannot slide sideways). In other words, nonholonomic constraints limit the system mobility by constraining the path that the robot can take from an initial pose to a final pose. From a control perspective, it has been shown that nonholonomic systems cannot be stabilized with continuous, static state feedback [95]. This is known as Brockett's condition.

---

1 To read about nonholonomic mechanical systems, one may consult [94].

4.2   Nonholonomic Kinematics

We will first only consider the kinematic equation 4.1a and design a velocity-level control law by treating η_i as the control input. As in previous chapters, we will make use of the basic formation acquisition control term formulated in Section 2.1.

4.2.1   Control Design

Since 4.1a is a *single-integrator-like* equation, we will decompose it as follows

$$\begin{bmatrix} \dot{x}_{ci} \\ \dot{y}_{ci} \end{bmatrix} = \begin{bmatrix} v_i \cos \theta_i \\ v_i \sin \theta_i \end{bmatrix} =: \begin{bmatrix} \bar{u}_{ix} \\ \bar{u}_{iy} \end{bmatrix} \tag{4.3a}$$

$$\dot{\theta}_i = \omega_i \tag{4.3b}$$

where $\bar{u}_{ix}, \bar{u}_{iy}$ are the velocities of point C_i in the x and y directions, respectively. If we could directly specify these velocities, then (4.3a) is identical to (2.1) and we just set $\bar{u}_i := [\bar{u}_{ix}, \bar{u}_{iy}]$ to (2.20). Therefore, the problem becomes to simply solve the algebraic equations

$$v_i \cos \theta_i = \bar{u}_{ix} \tag{4.4a}$$

$$v_i \sin \theta_i = \bar{u}_{iy} \tag{4.4b}$$

for v_i and θ_i where $\bar{u}_{ix}, \bar{u}_{iy}$ are given by the right-hand side of (2.20). If we multiply the top (resp., bottom) equation by $\cos \theta_i$ (resp., $\sin \theta_i$) and add them up, we obtain

$$v_i = \bar{u}_{ix} \cos \theta_i + \bar{u}_{iy} \sin \theta_i. \tag{4.5}$$

If we divide (4.4b) by (4.4a), we get

$$\theta_i = \text{atan2}(\bar{u}_{iy}, \bar{u}_{ix}). \tag{4.6}$$

Now, since we cannot directly specify θ_i, we let θ_{di} represent the desired value for θ_i and set it to the right-hand side of (4.6).[2] If $\tilde{\theta}_i = \theta_i - \theta_{di}$ is the orientation error, we have that

$$\dot{\tilde{\theta}}_i = \omega_i - \dot{\theta}_{di} \tag{4.7}$$

where

$$\dot{\theta}_{di} = \begin{cases} 0, & \text{if } \bar{u}_{ix} = \bar{u}_{iy} = 0 \\ \dfrac{\bar{u}_{ix}}{\bar{u}_{ix}^2 + \bar{u}_{iy}^2} \dot{\bar{u}}_{iy} - \dfrac{\bar{u}_{iy}}{\bar{u}_{ix}^2 + \bar{u}_{iy}^2} \dot{\bar{u}}_{ix}, & \text{otherwise} \end{cases}$$

---

2 Since the atan2 function is not defined when both its arguments are zero, we set $\theta_{di} = 0$ if $\bar{u}_{iy} = \bar{u}_{ix} = 0$.

and

$$\dot{\overline{u}}_i = \begin{bmatrix} \dot{\overline{u}}_{ix} \\ \dot{\overline{u}}_{iy} \end{bmatrix} = -k_v \sum_{j \in \mathcal{N}_i(E^*)} (z_{ij} + 2\tilde{q}_{ij}\tilde{q}_{ij}^\top)(\overline{u}_i - \overline{u}_j).$$

Based on (4.7), we can design

$$\omega_i = -k_\theta \tilde{\theta}_i + \dot{\theta}_{di} \tag{4.8}$$

to make $\tilde{\theta}_i = 0$ exponentially stable.

4.2.2 Simulation Results

The simulation of the kinematic control law given by (4.5) and (4.8) consisted of five vehicles forming the regular pentagon in Figure 4.2, where $q_1^* = [0, 0.2]$, $q_2^* = [-0.2s_1, 0.2c_1]$, $q_3^* = [-0.2s_2, -0.2c_2]$, $q_4^* = [0.2s_2, -0.2c_2]$, $q_5^* = [0.2s_1, 0.2c_1]$, $s_1 = \sin\frac{2\pi}{5}$, $s_2 = \sin\frac{4\pi}{5}$, $c_1 = \cos\frac{2\pi}{5}$, and $c_2 = \cos\frac{\pi}{5}$.

The initial pose of each vehicle was set to

$$p_1(0) = [0.0188 \text{ m}, 0.277 \text{ m}, 0.052 \text{ rad}]$$

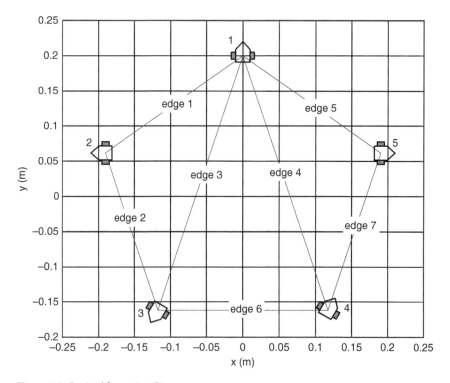

Figure 4.2 Desired formation F^*.

$p_2(0) = [-0.0769 \text{ m}, 0.1824 \text{ m}, -1 \text{ rad}]$

$p_3(0) = [0.01 \text{ m}, -0.0283 \text{ m}, -1.2 \text{ rad}]$

$p_4(0) = [0.2 \text{ m}, -0.2023 \text{ m}, 1 \text{ rad}]$

$p_5(0) = [0.3147 \text{ m}, 0.0396 \text{ m}, 0.1242 \text{ rad}].$

The control gains were chosen as $k_v = 5$ and $k_\theta = 1$.

The trajectory of the pose of point C_i for each vehicle is shown in Figure 4.3. Notice that the final orientation of each vehicle is 0 rad since $\bar{u}_i$ becomes zero when the formation is acquired and therefore the right-hand side of (4.6) also becomes zero.[3] The distance and orientation errors are depicted in Figure 4.4 while the control inputs are given in Figure 4.5.

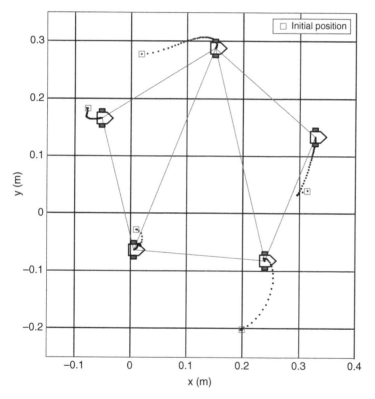

Figure 4.3 Trajectory of the poses $p_i(t), i = 1, \dots, 5$.

---

3 If it is necessary that each vehicle have some nonzero orientation upon acquiring the formation (e.g., in order to position a sensor or end-effector), a simple solution is to switch θ_{di} to the desired orientation after the distance errors converge to zero.

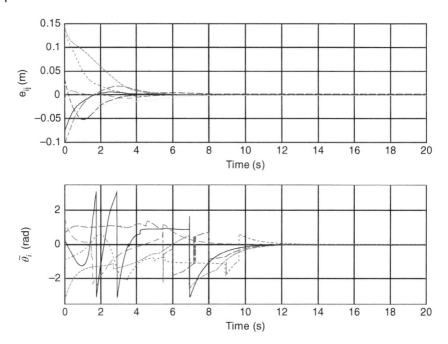

Figure 4.4 Distance errors $e_{ij}(t)$, $(i, j) \in E^*$ (top) and orientation errors $\tilde{\theta}_i(t)$, $i = 1, \ldots, 5$ (bottom).

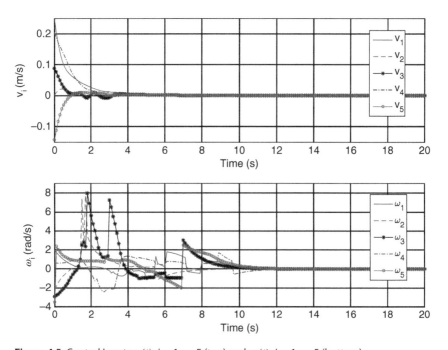

Figure 4.5 Control inputs $v_i(t)$, $i = 1, \ldots 5$ (top) and $\omega_i(t)$, $i = 1, \ldots 5$ (bottom).

4.3   Holonomic Dynamics

Here, we use a trick that bypasses the nonholonomic constraint present in (4.1) and allows us to treat the robot as an Euler–Lagrange system. To this end, we define the "hand" position for the ith robot as the point that lies a distance L_i along the X_i axis from point C_i (see point H_i in Figure 4.1). The hand position q_i is then given by

$$q_i = \begin{bmatrix} x_i \\ y_i \end{bmatrix} = \begin{bmatrix} x_{ci} \\ y_{ci} \end{bmatrix} + L_i \begin{bmatrix} \cos \theta_i \\ \sin \theta_i \end{bmatrix}. \tag{4.9}$$

In practice, the hand position could represent a point of interest on the robot such as an end-effector or a sensor.

The advantage of using (4.9) as the point to be controlled is that its kinematics are holonomic for any $L_i \neq 0$. Specifically, from (4.1a), (4.2), and (4.9), we have that

$$\eta_i = J(\theta_i)\dot{q}_i \tag{4.10}$$

where

$$J(\theta_i) = \begin{bmatrix} \cos \theta_i & \sin \theta_i \\ -\sin \theta_i & \cos \theta_i \\ \dfrac{}{L_i} & \dfrac{}{L_i} \end{bmatrix}, \tag{4.11}$$

which is invertible for $L_i \neq 0$.[4] The trade-off for this simplification is that we are no longer controlling the robot per se. Rather, we are controlling point H_i and the robot center of mass could end up anywhere on a circle of radius L_i around H_i.

Taking the time derivative of (4.10) and pre-multiplying the resulting equation by $\overline{M}_i$, we obtain

$$\overline{\tau}_i - \overline{D}_i J(\theta_i)\dot{q}_i = \overline{M}_i \dot{J}(\theta_i)\dot{q}_i + \overline{M}_i J(\theta_i)\ddot{q}_i \tag{4.12}$$

where (4.1b) and (4.10) were used. Now, pre-multiplying (4.12) by $J^{\mathsf{T}}(\theta_i)$, we arrive at the following Euler–Lagrange-like dynamic model

$$M_i(\theta_i)\ddot{q}_i + C_i(\theta_i, \dot{\theta}_i)\dot{q}_i + D_i(\theta_i)\dot{q}_i = \tau_i \tag{4.13}$$

where

$$M_i = J^{\mathsf{T}}(\theta_i)\overline{M}_i J(\theta_i), \quad C_i = J^{\mathsf{T}}(\theta_i)\overline{M}_i \dot{J}(\theta_i), \quad D_i = J^{\mathsf{T}}(\theta_i)\overline{D}_i J(\theta_i),$$

$$\text{and } \tau_i = J^{\mathsf{T}}(\theta_i)\overline{\tau}_i. \tag{4.14}$$

The expressions for the mass matrix $M_i(\theta_i)$ and the Coriolis/centripetal matrix $C_i(\theta_i, \dot{\theta}_i)$ are given in Appendix D.

---

4 Note that by rewriting (4.10) as $\dot{q}_i = J^{-1}(\theta_i)\eta_i$ and assuming η_i to be a velocity-level control input, one can set $\eta_i = J(\theta_i)u_i$ to recover the single-integrator model in (2.1).

The transformed dynamics (4.13) satisfy the following properties, which can be easily verified from the expressions in Appendix D. These properties will prove useful during the subsequent control design and stability analysis.

Property 4.1 The mass matrix is symmetric and positive definite, and

$$m_{i1}\|\mu\|^2 \le \mu^T M_i(\theta_i)\mu \le m_{i2}\|\mu\|^2 \quad \forall \mu \in \mathbb{R}^2 \tag{4.15}$$

where $m_{i2} > m_{i1} > 0$ are constants.

Property 4.2 The mass and Coriolis/centripetal matrices have the following skew-symmetric relationship:

$$\mu^T \left(\frac{1}{2}\dot{M}_i(\theta_i) - C_i(\theta_i, \dot{\theta}_i)\right)\mu = 0 \quad \forall \mu \in \mathbb{R}^2. \tag{4.16}$$

Property 4.3 The dynamics are linear in the parameters in the sense that

$$M_i(\theta_i)\dot{\mu} + C_i(\theta_i, \dot{\theta}_i)\mu + D_i(\theta_i)\dot{q}_i = Y_i(\theta_i, \dot{\theta}_i, \dot{q}_i, \mu, \dot{\mu})\phi_i \quad \forall \mu \in \mathbb{R}^2 \tag{4.17}$$

where

$$\phi_i = [m_i, \bar{I}_i/L_i^2, [\overline{D}_i]_{11}, [\overline{D}_i]_{12}/L_i, [\overline{D}_i]_{21}/L_i, [\overline{D}_i]_{22}/L_i^2] \tag{4.18}$$

is the constant parameter vector, $Y_i \in \mathbb{R}^{2\times6}$ is the regression matrix, and $[\cdot]_{ij}$ denotes the ijth component of the matrix.

4.3.1 Model-Based Control

In this section, we assume model (4.19) is exactly known for each of the n vehicles. We begin by rewriting (4.19) as

$$\dot{q}_i = v_i \tag{4.19a}$$

$$M_i(\theta_i)\dot{v}_i = \tau_i - C_i(\theta_i, \dot{\theta}_i)v_i - D_i(\theta_i)v_i \tag{4.19b}$$

where $v_i \in \mathbb{R}^2$ represents the hand velocity of the ith robotic vehicle relative to $\{X_0, Y_0\}$. Comparing (4.3.1) with (3.1), one can see that the double-integrator model is a simplified version of (4.3.1).

Due to this similarity, we use a Lyapunov function candidate akin to the one in (3.5). Namely, we introduce the function

$$W_m(e, s) = W(e) + \frac{1}{2}s^T M(\theta)s \tag{4.20}$$

where W was defined in (2.10), s was defined in (3.4)[5], $\theta = [\theta_1, \dots, \theta_n]$, and

$$M(\theta) = \text{diag}(M_1(\theta_1), \dots, M_n(\theta_n)).$$

---

5 In this chapter, the vectors v, v_f, and s have dimension $2n$ since the motion is planar.

Note that (4.20) is positive definite with respect to s because of Property 4.1. After taking the time derivative of (4.20), we obtain

$$\dot{W}_m = z^T R(\tilde{q})v + \frac{1}{2}s^T \dot{M}(\theta)s + s^T M(\theta)\dot{s}$$

$$= z^T R(\tilde{q})(s + v_f) + \frac{1}{2}s^T \dot{M}(\theta)s$$

$$+ s^T(u - C(\theta, \dot{\theta})\dot{q} - D(\theta)\dot{q} - M(\theta)\dot{v}_f)$$

$$= z^T R(\tilde{q})v_f + s^T(u - C(\theta, \dot{\theta})v_f - D(\theta)\dot{q} - M(\theta)\dot{v}_f + R^T(\tilde{q})z) \quad (4.21)$$

where (4.13) and (4.16) were used,

$$C(\theta, \dot{\theta}) = \text{diag}(C_1(\theta_1, \dot{\theta}_1), ..., C_n(\theta_n, \dot{\theta}_n)), \quad D(\theta) = \text{diag}(D_1(\theta_1), ..., D_n(\theta_n)),$$

and $u = [u_1, ..., u_n] \in \mathbb{R}^{2n}$.

The control law that solves the formation acquisition problem is given in the following theorem.

Theorem 4.1 Given the formation $F(t) = (G^*, q(t))$, let the initial conditions be such that $(e(0), s(0)) \in \Omega_1 \cap \Omega_2 \cap \Omega_3$ where Ω_1 and Ω_2 were defined in (2.14) and Ω_3 was defined in (3.7). Then, the model-based control

$$\tau = -k_a s + C(\theta, \dot{\theta})v_f + D(\theta)\dot{q} + M(\theta)\dot{v}_f - R^T(\tilde{q})z \quad (4.22)$$

where v_f is defined in (3.9): (i) renders $(e, s) = 0$ exponentially stable and ensures that (1.26) is satisfied, and (ii) ensures all other system signals remain bounded during closed-loop operation.

Proof: The proof of part (i) is very similar to the proof of Theorem 3.1 so some details are omitted. Substituting (4.22) into (4.21) leads to

$$\dot{W}_m = -k_v z^T R(\tilde{q})R^T(\tilde{q})z - k_a s^T s. \quad (4.23)$$

From (4.15) and (4.20), we have that

$$\dot{W}_m \leq -k_v \lambda_{\min}(R(\tilde{q})R^T(\tilde{q}))z^T z - k_a s^T s$$

$$\leq -\min\{k_v \lambda_{\min}(RR^T), k_a\}(\|z\|^2 + \|s\|^2)$$

$$\leq -\beta W_m \quad (4.24)$$

for $e(0) \in \Omega_1$, where

$$\beta = \frac{\min\{k_v \lambda_{\min}(RR^T), k_a\}}{\max\{1/4, 1/2 \max_i\{m_{i2}\}\}} > 0. \quad (4.25)$$

The rest follows as in the proof of Theorem 3.1.

Because we use a transformed model in the control design, it is important to show that all system signals (including "internal" signals) are bounded.

Since $(e, s) = 0$ is exponentially stable, we know $e(t), s(t) \in \mathcal{L}_\infty$, from which we can show $\tilde{q}(t), z(t) \in \mathcal{L}_\infty$ using (2.6) and (2.9). From (3.9) and (2.15), we then know $v_f(t) \in \mathcal{L}_\infty$. From (3.4), we can state that $v(t) \in \mathcal{L}_\infty$. From (3.12), we know $\dot{v}_f(t) \in \mathcal{L}_\infty$. Now, since e and s converge to zero *exponentially*, so does v (see (3.4) and (3.9)), therefore $q(t) = \int v(t)dt \in \mathcal{L}_\infty$ (see Appendix B). From (4.22), we conclude that $u(t) \in \mathcal{L}_\infty$. From (4.11), we can see that $J^{-1}(t) \in \mathcal{L}_\infty$ since all elements of the matrix are trigonometric terms. Therefore, we know from (4.14) that $\bar{u}_i(t) \in \mathcal{L}_\infty$. From (4.13) and (4.15), we have that $\ddot{q}(t) \in \mathcal{L}_\infty$. Since $J(t) \in \mathcal{L}_\infty$, we know $\eta_i(t) \in \mathcal{L}_\infty$ from (4.10). Then, from (4.1a) and (4.2), we have that $\dot{p}_{ci}(t) \in \mathcal{L}_\infty$. From (4.1b), it is obvious that $\dot{\eta}_i(t) \in \mathcal{L}_\infty$. Since $\dot{q}_i$ converges to zero exponentially, we know from (4.10) that η_i converges to zero exponentially. This in turn tells us that $\dot{p}_{ci}$, and therefore $\dot{\theta}_i$, converge to zero exponentially from (4.1a). As a result, $\theta_i(t) \in \mathcal{L}_\infty$. Finally, from (4.9), we know that $x_{ci}(t), y_{ci}(t) \in \mathcal{L}_\infty$. This concludes the proof of part (ii). $\square$

A comparison of (4.22) with (3.8) shows that the extra terms in (4.22) are used to cancel the dynamic terms that appear in (4.21), i.e., to feedback linearize the system. As a result, the right-hand sides of (4.23) and (3.10) are identical.

The ith control input is given by

$$\tau_i = (D_i(\theta_i) - k_v I_2)v_i + [C_i(\theta_i, \dot{\theta}_i) - (k_v k_q + 1)I_2] \sum_{j \in \mathcal{N}_i(E^*)} \tilde{q}_{ij} z_{ij}$$
$$- k_q M_i(\theta_i) \sum_{j \in \mathcal{N}_i(E^*)} (z_{ij}I_2 + 2\tilde{q}_{ij}\tilde{q}_{ij}^\mathsf{T})\tilde{v}_{ij} \tag{4.26}$$

where $\tilde{v}_{ij}$ was defined in (3.16). In comparison to (3.15), the control input for the ith vehicle is also a function of its own heading angle and rate, which can be measured with onboard sensors. Notice that the controller does not depend on the hand position q_i or the center of mass position $[x_{ci}, y_{ci}]$.

4.3.2 Adaptive Control

Here we consider the more realistic case where the parameters in (4.18) are subject to uncertainty and therefore their values are unknown to the designer.

First, by making use of Property 4.3, (4.21) can be rewritten as

$$\dot{W}_m = z^\mathsf{T} R(\tilde{q})v_f + s^\mathsf{T}(u - Y(\theta, \dot{\theta}, \dot{q}, v_f, \dot{v}_f)\phi + R^\mathsf{T}(\tilde{q})z), \tag{4.27}$$

where

$$Y(\theta, \dot{\theta}, \dot{q}, v_f, \dot{v}_f) = Y_1(\theta_1, \dot{\theta}_1, \dot{q}_1, v_{f1}, \dot{v}_{f1}) \oplus \cdots \oplus Y_n(\theta_n, \dot{\theta}_n, \dot{q}_n, v_{fn}, \dot{v}_{fn})),$$

$\oplus$ represents the matrix direct sum (see Appendix A), and $\phi = [\phi_1, ..., \phi_n] \in \mathbb{R}^{6n}$. Likewise, the model-based controller (4.22) can be expressed as

$$\tau = -k_a s + Y(\theta, \dot{\theta}, \dot{q}, v_f, \dot{v}_f)\phi - R^\mathsf{T}(\tilde{q})z. \tag{4.28}$$

We now have the constraint that the parameter vector ϕ is unknown and cannot be used in the control law. Therefore, the formation controller will include a *dynamic estimate* of each ϕ_i, whose adaptation law will be part of the control design. To this end, let $\hat{\phi}_i(t) \in \mathbb{R}^6$ be the ith parameter estimate and define the corresponding parameter estimation error as

$$\tilde{\phi}_i = \hat{\phi}_i - \phi_i .\tag{4.29}$$

To solve the problem, we use the (indirect) adaptive control

$$\tau = -k_a s + Y(\theta, \dot{\theta}, \dot{q}, v_f, \dot{v}_f)\hat{\phi} - R^{\mathsf{T}}(\tilde{q})z \tag{4.30a}$$

$$\dot{\hat{\phi}} = -\Gamma Y^{\mathsf{T}}(\theta, \dot{\theta}, \dot{q}, v_f, \dot{v}_f)s \tag{4.30b}$$

where v_f was defined in (3.9), $\hat{\phi} = [\hat{\phi}_1, ..., \hat{\phi}_n]$, and $\Gamma \in \mathbb{R}^{6n \times 6n}$ is constant, diagonal, and positive definite. This is a certainty equivalence-type control law since ϕ is simply replaced by the estimate $\hat{\phi}(t)$ that comes from the adaptation law (4.30b). The following theorem delineates the stability result we obtain with (4.30).

Theorem 4.2 Let $\tilde{\phi} = [\tilde{\phi}_1, ..., \tilde{\phi}_n]$, $\xi = [e, s, \tilde{\phi}]$, and the initial conditions satisfy $\xi(0) \in S := (\Omega_1 \cap \Omega_2 \cap \Omega_3) \times \mathbb{R}^{6n}$ where Ω_1 and Ω_2 were defined in (2.14) and Ω_3 was defined in (3.7). Then, the adaptive control (4.30) ensures that (1.26) is met.

Proof: Define the Lyapunov function candidate

$$W_a(e, s, \tilde{\phi}) = W_m(e, s) + \frac{1}{2}\tilde{\phi}^{\mathsf{T}}\Gamma^{-1}\tilde{\phi} \tag{4.31}$$

where W_m was given in (4.20). Differentiating (4.31) and then substituting (4.30a) yields

$$\dot{W}_a = -k_v z^{\mathsf{T}} R R^{\mathsf{T}} z - k_a s^{\mathsf{T}} s + s^{\mathsf{T}} Y\tilde{\phi} - \tilde{\phi}^{\mathsf{T}}\Gamma^{-1}\dot{\hat{\phi}} \tag{4.32}$$

where (4.27) was used. Now, after substituting (4.30b) into (4.32), we obtain

$$\dot{W}_a = -k_v z^{\mathsf{T}} R R^{\mathsf{T}} z - k_a s^{\mathsf{T}} s. \tag{4.33}$$

By following the arguments originally set forth in the proof of Theorem 2.1, we can state

$$\dot{W}_a \leq -k_v \lambda_{\min}(R R^{\mathsf{T}})\|z\|^2 - k_a\|s\|^2 \quad \text{for } e(0) \in \Omega_1. \tag{4.34}$$

From (4.31) and (4.34), we know that $z(t), s(t), \tilde{\phi}(t) \in \mathcal{L}_\infty$. From (2.6) and (2.8), we then know $\tilde{q}(t), e(t) \in \mathcal{L}_\infty$. From (3.9), we know $v_f(t) \in \mathcal{L}_\infty$; thus, $v(t) \in \mathcal{L}_\infty$ from (3.4). From (3.14) and (3.12), we know $\dot{z}(t), \dot{v}_f(t) \in \mathcal{L}_\infty$. Since $J(t) \in \mathcal{L}_\infty$, we know $\eta_i(t) \in \mathcal{L}_\infty$ from (4.10). From (4.1a) and (4.2), we have that $\dot{p}_i(t) \in \mathcal{L}_\infty$. From (4.30a) and the fact that θ appears only through trigonometric functions, we know $\tau(t) \in \mathcal{L}_\infty$. Then, since $J^{-1}(t) \in \mathcal{L}_\infty$, we have that $\overline{\tau}_i(t) \in \mathcal{L}_\infty$ from (4.14). From (4.13) and (4.15), we know $\ddot{q}(t) \in \mathcal{L}_\infty$. From (4.1b), we know $\dot{\eta}_i(t) \in \mathcal{L}_\infty$.

It follows from (4.34) that $z(t) \in \mathcal{L}_2$ for $e(0) \in \Omega_1$. Since we previously showed that $z(t), \dot{z}(t) \in \mathcal{L}_\infty$, we can invoke Theorem C.3 to conclude that $z(t) \to 0$ as $t \to \infty$ for $e(0) \in \Omega_1$. Therefore, we know from (2.9) that $e(t) \to 0$ as $t \to \infty$ for $e(0) \in \Omega_1$. From this point, we can again proceed as in the proof of Theorem 3.1 to claim $F(t) \to \mathrm{Iso}(F^*)$ as $t \to \infty$ for $\xi(0) \in S$. $\qquad\square$

4.3.3 Simulation Results

A five-vehicle simulation was conducted using the following parameters: $m_i = 3.6$ kg, $\overline{I}_i = 0.0405$ kg-m^2, $\overline{D}_i = \mathrm{diag}(0.3$ kg/s, 0.004 kg-m^2/s$)$, and $L_i = 0.15$ m for $i = 1, \dots, 5$. The simulation consisted of applying control law (4.30) to (4.1) using the fact that $\overline{\tau}_i = J_{i\tau i}^{-T}$ from (4.14). The desired formation was the regular convex pentagon described in Section 2.6.1.

The initial position of the ith vehicle, $q_i(0)$, was randomly chosen as a perturbation about q_i^* while its initial orientation, $\theta_i(0)$, was randomly set to a value between 0 and 2π. The initial position of each vehicle's mass center $[x_{ci}, y_{ci}]$ was then obtained from (4.9). The initial translational and angular speed of each vehicle were set to $v_i(0) = [0, -0.0393, 0.4816, -0.3436, 0.3555]$ m/s and $\dot{\theta}_i(0) = [0.1448, -0.1237, -0.3091, -0.0717, -0.0180]$ rad/s, respectively. The initial conditions for the parameter estimate vector was $\hat{\phi}(0) = 0$. The control and adaptation gains were set to $k_v = 1$, $k_a = 2$, and $\Gamma = I_{30}$.

Figure 4.6 shows the trajectories of the robots' hand position $q_i(t), i = 1, \dots, 5$ forming the desired shape, while Figure 4.7 shows the distance errors $e_{ij}(t)$, $i, j \in V^*$ converging to zero. Notice that the vehicle orientations are rather random upon reaching the final position. This is because the controller is based on the hand position, which is a *point*, rather than on the vehicle position and orientation. The actual control inputs applied to (4.1) are depicted in Figure 4.8. As an example of the behavior of the parameter estimates, the parameter estimates for vehicle 1, $\hat{\phi}_1(t)$, are shown in Figure 4.9. The fourth and fifth components of $\hat{\phi}_1$ converge to zero since they are related to $[\overline{D}_1]_{12}$ and $[\overline{D}_1]_{21}$, respectively, which were set to zero in the simulation. The parameter estimates for all the other vehicles also converged to constants as expected.

4.4 Notes and References

Some work in the literature has accounted for the vehicle kinematics and dynamics during the design of coordination controllers for multi-robot systems. For nonholonomic models, results can be divided into two categories: the

Figure 4.6 Trajectory of the hand positions $q_i(t)$, $i = 1, \ldots, 5$.

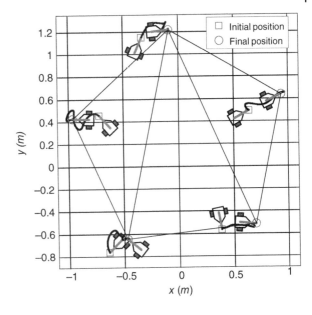

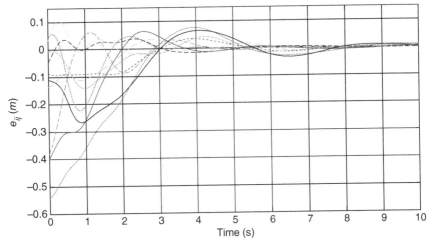

Figure 4.7 Distance errors $e_{ij}(t)$, $i, j \in V^*$.

purely kinematic model where the control inputs are at the velocity level and the dynamic model where the inputs are at the actuator level. Examples of work based on the kinematic model are the following. A class of simple control laws for assembling and coordinating the motions of nonholonomic vehicle formations was discussed in [31]. In [96], the nonholonomic kinematics was used to design a formation maneuvering controller and experimental results

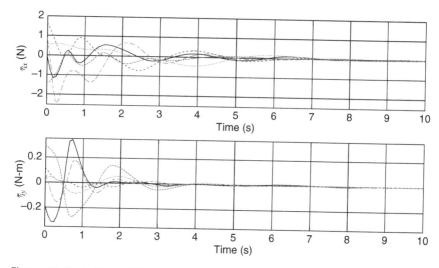

Figure 4.8 Control inputs $\overline{\tau}_i(t)$, $i = 1, \ldots, 5$.

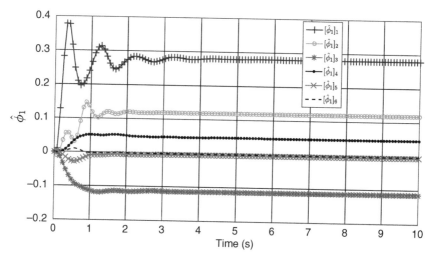

Figure 4.9 Parameter estimates for vehicle 1, $\hat{\phi}_1(t)$.

were presented for three-wheeled mobile robots. Unicycle robot kinematics were used in [97, 98] for designing formation maneuvering controllers. Vision-based control laws for parallel and balanced circular formations using a consensus approach were developed in [99]. In [100], a leader–follower-type solution was presented for the formation maneuvering problem where the inter-vehicle interactions are modeled by a spanning tree graph. In [101], a sliding mode controller based on a nonholonomic kinematic model was proposed to stabilize the inter-robot distances in a cyclic polygon formation. In [102, 103], the rendezvous and formation acquisition problems for unicycle kinematic agents were solved using a discontinuous, time-invariant control law.

For the case of nonholonomic dynamics, the model of a unicycle robot was used in [104] to design a formation control scheme that maintains the prescribed formation while avoiding obstacle and inter-vehicle collisions. In [105], a flocking and connectivity-preserving control algorithm was proposed using each robot's state and the heading angles of neighboring robots. In [2, 36], a class of coordination schemes, including aggregation, foraging, formation acquisition, and target interception controllers, were presented for holonomic and nonholonomic dynamics with uncertainty. The work in [106] introduced a receding-horizon, leader–follower control framework to solve the formation problem with a rapid error convergence rate.

Examples of work based on the holonomic dynamic model are the following. In [77], consensus-type controller–observers were formulated to allow a team of followers to track a dynamic leader whose motion is known by only a subset of the followers. A synchronization tracking controller was designed in [107] for the cooperative multi-robot system. A finite-time consensus tracking controller for leader–follower multi-robot systems was proposed in [108]. In [109], a robust adaptive formation controller was designed under the presence of parameter uncertainties in the system model. Under the assumption of functional uncertainties, [110] constructed a neural network controller that ensures the multi-robot system is synchronized with the motion of a dynamic target. In [111], a passive decomposition approach was used to decouple the solution of the formation acquisition and formation maneuvering problems. In [112], an adaptive neural network controller was introduced for formations of marine vessels with uncertain dynamics using the dynamic surface control technique. A formation acquisition and flocking-type controller was designed in [113] for a fleet of ships using the integrator backstepping technique. Other work that employed backstepping as a means of compensating for the robot dynamics during formation control includes [114–116]. The formation maneuvering of fully actuated marine vessels was studied in [117] using the passivity-based group coordination framework. In [118], a target interception scheme was developed using sliding mode control for vehicle dynamics subject to uncertainty and disturbances.

The material in this chapter is partly based on the work in [119, 120].

5

Experimentation

Experimental validations of the formation controllers from the previous chapters were conducted to illustrate how the algorithms perform on an actual robotic platform. The physical implementation of the controllers and the results of each experiment are reported in this chapter.

5.1 Experimental Platform

The experiments were conducted using three of the customized Traxxas E-Maxx electric UGVs shown in Figure 5.1. The UGV is about 48 cm in length and 38 cm in width, and is powered by two 7 C-cell battery packs for a maximum voltage of 16 V at full charge.

Several modifications were performed to the stock vehicle. The UGV's electronic speed controller was replaced with the RoboteQ SDC1130 H-bridge motor amplifier. Furthermore, one of the two DC motors was disconnected to reduce the vehicle's maximum acceleration and speed. Additional springs were placed in the suspension to help support the weight of the sensing platform. Finally, the UGV was equipped with the following sensor/communication/processing suite:

- A Bosch BNO055 inertial measurement unit (IMU) equipped with three-axis accelerometers/gyros and a magnetic compass. The BNO055 provides a sensor-fused output for heading (i.e., tilt compensated).
- A US Optical 512 pulses per revolution (ppr) incremental encoder coupled to the UGV's transmission shaft to provide vehicle displacement measurement.
- An ARM-based mBed microprocessor to perform sensor, control, and communication operations.
- A Digi xBee PRO wireless serial modem set to a baud rate of 115, 200 bps for inter-vehicle communication.
- A radio-controlled (RC) triggered relay board that allows for switching between autonomous operation (via the mBed) and teleoperation (via a Futaba RC transmitter) of the vehicle.

Formation Control of Multi-Agent Systems: A Graph Rigidity Approach, First Edition.
Marcio de Queiroz, Xiaoyu Cai, and Matthew Feemster.
© 2019 John Wiley & Sons Ltd. Published 2019 by John Wiley & Sons Ltd.
Companion website: www.wiley.com/go/dequeiroz/formation_control

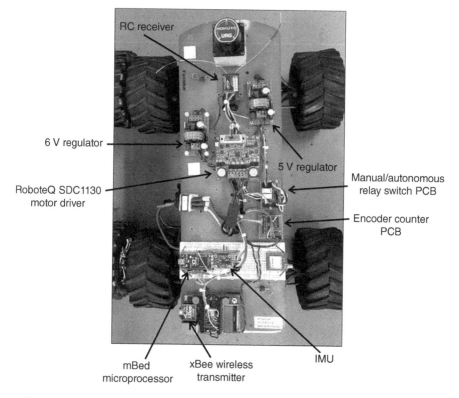

Figure 5.1 Top view of the experimental UGV platform.

The encoder was attached to a take-off point of the transmission (see Figure 5.2). Thus, the gear ratio of the vehicle magnified the 512 ppr of the encoder shaft to approximately 10, 900 ppr of the vehicle's wheel. This led to a linear displacement accuracy of 0.44 mm. This extremely accurate displacement measurement combined with the fused compass reading provided by the IMU gave an extremely reliable odometry measurement for calculation of the UGV position using only onboard sensors. That is, GPS was not used to obtain position information. Finally, each vehicle's velocity was obtained from a backward difference algorithm applied to the position measurement. This signal was then passed through a low-pass digital filter with cut-off frequency of approximately 5 Hz.

The centralized scheme shown in Figure 5.3 was adopted for implementation of the formation control algorithms. Each vehicle transmitted its global position (x_i, y_i) and heading θ_i signals via the xBee serial modem to a laptop (Microsoft Surface Book, 2.6 GHz) running Mathwork's MATLAB. The control schemes were coded and computed in MATLAB, and each UGV's commanded steering angle δ_i and drive-wheel motor voltage V_i were transmitted back to the

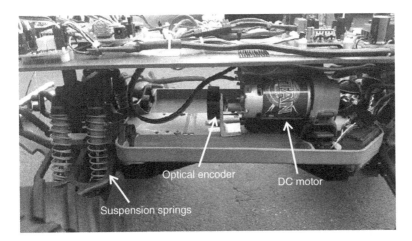

Figure 5.2 Side view of the UGV.

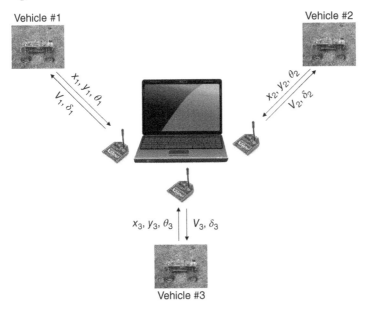

Figure 5.3 Experimental control and communication scheme.

vehicle. The mBed processor onboard each UGV was used to process the sensor measurements and apply the actuator commands. Specifically, the processor performed the following sequence of functions:

1. Processed the wheel displacement and vehicle heading measurements by communicating via a SPI bus to the LS766 chip that reads the optical encoder, and via an I2C bus to the IMU.

2. Transmitted the measurements to the laptop via the xBee modem.
3. Received the actuator commands from the laptop via the xBee modem.
4. Generated the corresponding PWM signals for the drive-wheel motor and steering servo.

Admittedly, the chosen implementation scheme does not conform directly with the definition of a distributed/decentralized system presented in Definition 1.1. However, the decision to use the centralized scheme was based solely on the limited performance capabilities of the existing onboard processing units, and was not due to any restrictions or assumptions imposed by the formation control algorithms themselves. With upgrades to the vehicle's hardware capabilities, a shift to a more decentralized implementation (where the control algorithms are implemented on the each vehicle) could be easily obtained. The arrangement of Figure 5.3 was chosen over a decentralized scheme for the following reasons:

- Since the formation control algorithms were initially simulated in the MATLAB software, all necessary control functions were readily available and did not need to be converted to C for implementation on the vehicle's ARM processor. That is, to preform the experiments we had to only replace the mathematical model of the vehicle with the hardware interface to the UGVs.
- The mBed processors utilized on the vehicles do not have a matrix library readily available. Therefore, conversion of the matrix calculations in the control algorithms is somewhat tedious and prone to errors.
- Due to the higher capable processor of the laptop computer, the control frequency increased to approximately 40 Hz for the centralized scheme from the 14Hz frequency when the entire formation control algorithm was implemented directly on the vehicle's mBed processor.
- Control gain tuning is greatly simplified as all control calculations are implemented via MATLAB script. Therefore, the need for attaching USB cables for compiling and downloading C programs to the mBed processors each time a control gain is changed was eliminated.
- Data logging and performance evaluation is simplified as all vehicle states are recorded by MATLAB as the control program executes.

5.2   Vehicle Equations of Motion

The UGV shown in Figure 5.1 is a nonholonomic car-like robot [121] whose equations of motion differ from the unicycle robot described in Section 4.1.[1] In the following, the kinematic and dynamic models of the UGV are derived.

---

1 Since the controller of Section 4.2 is based on the unicycle model, it was not experimentally tested.

Figure 5.4 Schematic of the
experimental UGV.

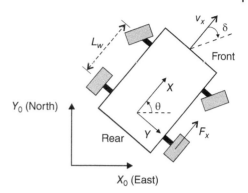

Figure 5.4 shows a schematic of the UGV where $\{X_0, Y_0\}$ is Earth-fixed reference frame and $\{X, Y\}$ is the body-fixed reference frame located at the midpoint of the rear axle. The UGV is rear-wheel driven by a DC motor and front-wheel steered by a servo motor like a regular car. In the derivation of the UGV model, we make the following assumptions:

- The mass of the vehicle is evenly distributed.
- The two wheels on each axle (front and rear) can be modeled as a single wheel located at the midpoint of each axle.
- The wheels roll without slipping when in contact with the ground (nonholonomic constraint); hence, the lateral velocity of each wheel is zero.
- The half tread of the vehicle tires is small, i.e., the tires are relatively smooth.
- The aerodynamic drag is negligible due to the vehicle's low-speed operation.

The kinematic equations for the vehicle center of mass can be modeled by [121]

$$\begin{bmatrix} \dot{x} \\ \dot{y} \\ \dot{\theta} \end{bmatrix} = \begin{bmatrix} \cos\theta \\ \sin\theta \\ \dfrac{1}{L_w}\tan\delta \end{bmatrix} v_x \tag{5.1}$$

where (x, y, θ) represent the position and orientation of $\{X, Y\}$ with respect to $\{X_0, Y_0\}$, δ is the front wheel steering angle, L_w is the distance between the front and rear axles, and v_x is the vehicle's longitudinal speed (i.e., in the X-axis direction).

The UGV longitudinal dynamics is described by the following equation

$$m\dot{v}_x = -b_v v_x - F_s \text{sgn}(v_x) + F_x \tag{5.2}$$

where m is the vehicle mass, b is the viscous friction coefficient between the wheels and the ground, F_s is the static friction, and F_x denotes the total longitudinal rear force. Since the rear wheels do not slip on the driving surface, the longitudinal rear force is given by [122]

$$F_x - \frac{\tau_m}{r_w} \tag{5.3}$$

where r_w is the wheel radius and τ_m represents the output torque from the DC motor. Under the assumption that the DC motor's inductance is negligible, we have the following relationship between the motor input armature voltage V and its output torque

$$\tau_m = \frac{K_\tau}{R_a}(V - K_b\omega_m) \tag{5.4}$$

where R_a is the armature resistance, K_τ is the torque constant, K_b is the back-emf constant, and ω_m is the DC motor angular velocity given by

$$\omega_m = \frac{v_x}{r_w}. \tag{5.5}$$

After substituting (5.3–5.5) into (5.2), we obtain

$$m\dot{v}_x = -\left(b + \frac{K_\tau K_b}{r_w^2 R_a}\right)v_x - F_s\mathrm{sgn}(v_x) + \frac{K_\tau}{r_w R_a}V \tag{5.6}$$

Finally, from the time derivative of (5.1), we have that the translational dynamics for the UGV center of mass are governed by

$$\ddot{\zeta} = \begin{bmatrix} \ddot{x} \\ \ddot{y} \end{bmatrix} = \begin{bmatrix} -v_x\dot{\theta}\sin\theta + \dot{v}_x\cos\theta \\ v_x\dot{\theta}\cos\theta + \dot{v}_x\sin\theta \end{bmatrix}$$

$$= B \begin{bmatrix} \dfrac{v_x^2}{L_w}\tan\delta \\[2ex] -\left(b + \dfrac{K_\tau K_b}{r_w^2 R_a}\right)\dfrac{v_x}{m} - F_s\mathrm{sgn}(v_x) + \dfrac{K_\tau}{mr_w R_a}V. \end{bmatrix} \tag{5.7}$$

where the third equations of (5.1) and (5.6) were used, $\zeta = [x, y]$, and

$$B = \begin{bmatrix} -\sin\theta & \cos\theta \\ \cos\theta & \sin\theta \end{bmatrix}. \tag{5.8}$$

Notice that (5.8) is always invertible.

The values of the physical parameters in the above equations were experimentally identified to be

$$m = 5.54 \text{ kg} \qquad b_v = 1.105 \text{ N–s/m} \qquad F_s = 3.6 \text{ N}$$
$$L_w = 0.3556 \text{ m} \qquad r_w = 0.0762 \text{ m} \qquad R_a = 0.439 \text{ }\Omega \tag{5.9}$$
$$K_\tau = 0.0857 \text{ Nm/A} \qquad K_b = 0.0857 \text{ V–s/rad}$$

Figure 5.5 shows the experimentally determined total friction force acting on the UGV from which the values for b_v and F_s were obtained.

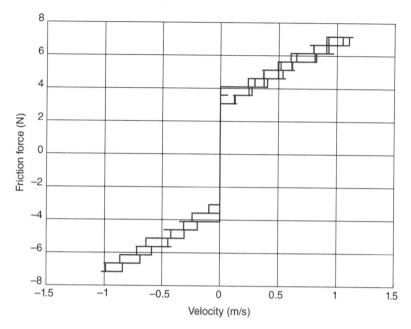

Figure 5.5 Friction force on the UGV.

5.3 Low-Level Control Design

Some (low-level) control input transformations had to be applied to (5.7) to enable the implementation of the (high-level) formation control algorithms. The primary purpose of the low-level controller is to calculate the correct steering angle δ and DC motor voltage V to support the formation control objective.

To this end, we first design the motor voltage and steering angle as

$$V = \frac{mr_w R_a}{K_\tau} \left[\left(b + \frac{K_\tau K_b}{r_w^2 R_a} \right) \frac{v_x}{m} + \frac{F_s}{m} \operatorname{sgn}(v_x) + u_v \right] \tag{5.10a}$$

$$\delta = \arctan \left(\frac{L_w}{v_x^2} u_\delta \right) \tag{5.10b}$$

where u_v, u_δ represent auxiliary control inputs to be specified soon. After substituting (5.10) into (5.7), we obtain

$$\ddot{\zeta} = B \begin{bmatrix} u_\delta \\ u_v \end{bmatrix} \tag{5.11}$$

Notice that (5.11) is a double-integrator-like model. Thus, if u^{DI} denotes any of the acceleration-level formation controllers from Chapter 3 for a single UGV (e.g., the right-hand side of (3.15) for the case of formation acquisition), then

$$\begin{bmatrix} u_\delta \\ u_v \end{bmatrix} = B^{-1} u^{DI} \tag{5.12}$$

applies the double integrator model-based control laws to the vehicles. The relationship in (5.12) is then used in (5.10) to implement the overall control system for a single UGV.

For the case of the single-integrator model, the output of the formation controller for a single UGV is a velocity command, which we denote here as u^{SI}. To quantify this velocity control objective, we define the following velocity error for a single UGV

$$e_v = u^{SI} - \dot{\zeta}. \tag{5.13}$$

After taking the time derivative of (5.13) and then neglecting the derivative of the velocity command, we obtain from (5.11) that

$$\dot{e}_v = -B \begin{bmatrix} u_v \\ u_\delta \end{bmatrix}. \tag{5.14}$$

Based on (5.14), the auxiliary control inputs are designed as

$$\begin{bmatrix} u_v \\ u_\delta \end{bmatrix} = B^{-1} \left(K_p e_v + K_I \int e_v(t) dt \right), \tag{5.15}$$

where $K_p, K_I \in \mathbb{R}^{2 \times 2}$ are diagonal, positive-definite, control gain matrices, yielding the exponentially stable closed-loop system

$$\dot{e}_v = -K_p e_v - K_I \int e_v(t) dt. \tag{5.16}$$

Note that (5.13) is essentially the velocity error defined in (3.4) with K_p in (5.15) playing a similar role to k_a in (3.8). That is, the main difference between (5.15) and the double-integrator-based controller in (3.8) is the absence of the "feedforward" terms $\dot{v}_f - R^\top(\tilde{q})z$ and the addition of an integral feedback term. The absence of the feedforward terms is compensated by using *high-gain* feedback.

To summarize the above discussion, the overall control system implemented on the UGVs for the purpose of applying the single-integrator model-based formation controllers consisted of (5.10), (5.13), (5.8), and (5.15) with u^{SI} set to a high-level control law from Chapter 2. For example, for formation acquisition, u^{SI} is given by the right-hand side of (2.20).

5.4 Experimental Results

A series of experiments was conducted to showcase the real-world performance of the formation control laws from the previous chapters. The control gains in the following experiments were selected by trial and error with the intent of achieving an *acceptable* performance in terms of the formation control objective being tested. That is, the gains were not exhaustively tuned for each control law to optimize the performance according to certain specifications. Although the meaning of "acceptable" is subjective, the following results demonstrate that our implementation approach was sufficient in producing formation performances that are well within what one would expect from control system practice. More importantly, the experimental results show that the control theory discussed in this book can be successfully implemented on an actual robotic platform.

The experimental trials were conducted in a parking lot located at (39.085817°, −76.589453°). As can be seen from Figure 5.6, the parking lot is aligned along a magnetic bearing of 32° and exhibits a slight uphill grade. Therefore, the introduction of the integral term in (5.15) was partially motivated by the desire to compensate for the gravity effect on the vehicles. Furthermore, since the parking lot exhibits various dips/valleys in numerous locations (though not severe), random fluctuations may be observed in the results depending upon the formation's location. Finally, due to measurement noise, sensor resolution, and quantization errors, the experimental variables will *not* necessarily approach the exact values predicted by the theory but rather will approach approximate values.

The basic formation acquisition component of the experiments consisted of a triangular formation with the desired inter-vehicle distances set to $d_{12} = d_{13} = d_{23} = 4$ m. The three vehicles were placed at rest in the following

Figure 5.6 Initial configuration of the UGVs during the experimental runs.

approximate initial positions and orientations

$$
\begin{bmatrix} x_1(0) \\ y_1(0) \\ \theta_1(0) \end{bmatrix} = \begin{bmatrix} 0 \text{ m} \\ 0 \text{ m} \\ 0° \end{bmatrix}, \begin{bmatrix} x_2(0) \\ y_2(0) \\ \theta_2(0) \end{bmatrix} = \begin{bmatrix} 0 \text{ m} \\ -1 \text{ m} \\ 0° \end{bmatrix}, \begin{bmatrix} x_3(0) \\ y_3(0) \\ \theta_3(0) \end{bmatrix} = \begin{bmatrix} -1 \text{ m} \\ -0 \text{ m} \\ 0° \end{bmatrix}.
\tag{5.17}
$$

The initial configuration of the UGVs is shown in Figure 5.6.

Each vehicle's global position was calculated via the following algorithm

$$
L_i[k] = r_w \alpha_i[k]
$$
$$
x_i[k] = x_i[k-1] + (L_i[k] - L_i[k-1]) \cos(\theta_i[k])
\tag{5.18}
$$
$$
y_i[k] = y_i[k-1] + (L_i[k] - L_i[k-1]) \sin(\theta_i[k])
$$

where $L_i[k]$ represents the longitudinal distance traveled in meters by the ith vehicle at the kth time sample, and $\alpha_i[k]$ is the wheel rotational displacement at the kth sample. Since an incremental optical encoder was utilized, $\alpha_i[0] = 0$ rad and the initial conditions for $x_i[0]$, $y_i[0]$, and $\theta_i[0]$ were selected according to (5.17). With the utilization of the incremental odometry approach in (5.18) to obtain the vehicles' global position, one can take advantage of the fact that the vehicles are in fact physically spaced further apart than what the sensor calculations indicate. This provides an invaluable asset during experimental tuning by avoiding potential vehicle collisions and resulting hardware damage during initial tuning trials. Specifically, it was observed that depending on initial placement of the vehicles, they would crisscross to achieve the inter-vehicle distance control objective. If the distance was too small, vehicle collision was likely to occur. Obviously, software safety measures could be invoked to stop the execution of the experimental trial when the distance approaches some minimum threshold. However, the premature termination of the control algorithm would not allow one to observe if formation convergence does indeed take place at a later time. Once the tuning process matured, the UGV's were then physically placed accurately to properly reflect the measurements presented.

The drive-wheel DC motor voltages in the following experiments were often amplitude-limited by software to restrict the acceleration and/or speed experienced by each UGV. This was done to prolong the usability of the UGVs and to avoid them reaching the end of the parking lot during the experimental run. Different saturation values were employed depending on the circumstances of each experiment.

Although the results shown next were taken from a single experimental run (after numerous tuning trials), the results were indeed repeatable when the experiment was re-run with the same control gains and initial conditions. The organization of the experimental results in the following sections closely follows the order in which the controllers were presented in the previous chapters.

5.4.1 Single Integrator: Formation Acquisition

The control gain in (2.15) was set to $k_v = 0.08$ while $K_P = 5I_2$ and $K_I = 0$ in (5.15) for all three vehicles. The position of each UGV as they formed the desired triangle is shown in Figure 5.7, where the markers denote the vehicle location at different instants of time during the interval $t \in [0, 40]$ s. The inter-vehicle distance errors defined in (2.6) are depicted in Figure 5.8, showing that the desired formation was acquired in less than 5 seconds. Notice from Figures 5.7 and 5.8 that the vehicles exhibit a small oscillatory motion near the final formation configuration. This is due to the fact that the formation acquisition objective is a *setpoint* control problem, therefore the vehicles are operating near zero velocity where static friction effects are prominent. Furthermore, the vehicles have noticeable backlash in the mechanical transmission. Since the encoder was collocated with the DC motor's drive shaft, these small position perturbations are measured by the encoder, but are often unobserved at the vehicle level due to the flexibility of the gear train.

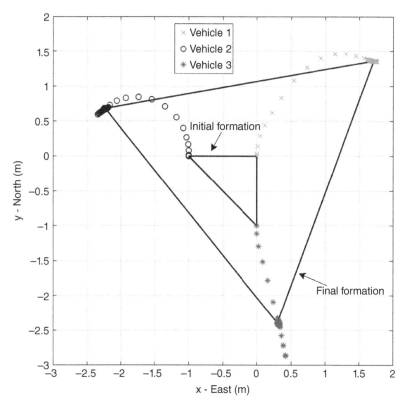

Figure 5.7 Single integrator: formation acquisition. UGV position trajectories $q_i(t) = [x_i(t), y_i(t)]$, $i = 1, 2, 3$.

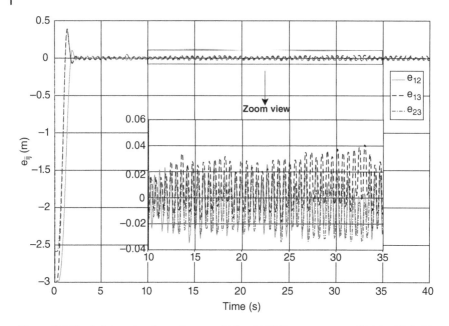

Figure 5.8 Single integrator: formation acquisition. UGV distance errors $e_{ij}(t)$, $i,j \in \{1,2,3\}$.

The control voltages applied to the each vehicle's drive-wheel DC motor is shown in Figure 5.9, and are well below the maximum battery voltage of 16 V. The constant switching observed in the voltages after the UGVs reach the desired formation is due to the static friction compensation term, $F_s \text{sgn}(v_x)$, in (5.10a). Due to imperfect, noisy measurements, v_x "chatters" about zero in the steady state and causes this term to constantly switch. The steering angle commands to the servo motors that orient the UGV front wheels are given in Figure 5.10, showing that they quickly became saturated at their maximum value of approximately $25°$. This saturation can be attributed to the vehicles being regulated to a setpoint position, which results in the vehicle longitudinal speed v_x approaching zero. As v_x decreases over time, the steering angle becomes larger according to (5.10b).

Figure 5.11 displays the x- and y-direction components of the velocity error (5.13) for each UGV. When tuning the control gain K_p, a balance was struck between acceptable steady-state performance and limiting the transient amplitude of the velocity errors.

5.4.2 Single Integrator: Formation Maneuvering

Two formation maneuvering experiments were conducted: one with formation translation only, and the other with translation and rotation. In both experiments, the UGVs started in the initial configuration of (5.17) while the control gains in (2.23) and (5.15) were set to $k_v = 0.08$, $K_p = 5I_2$, and $K_I = 0.5I_2$.

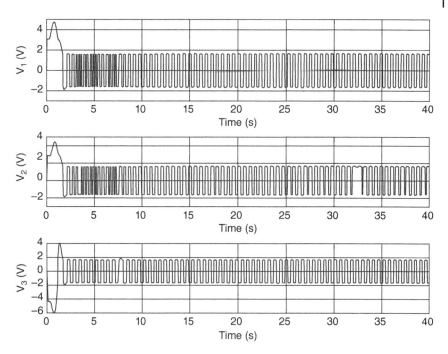

Figure 5.9 Single integrator: formation acquisition. Drive-wheel DC motor voltages $V_i(t)$, $i = 1, 2, 3$.

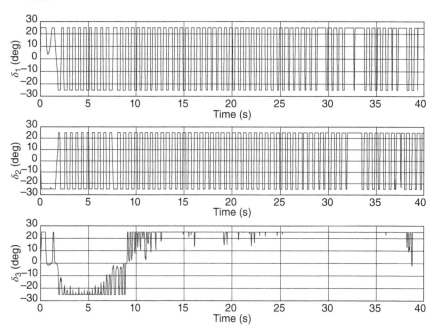

Figure 5.10 Single integrator: formation acquisition. Steering angle commands $\delta_i(t)$, $i = 1, 2, 3$.

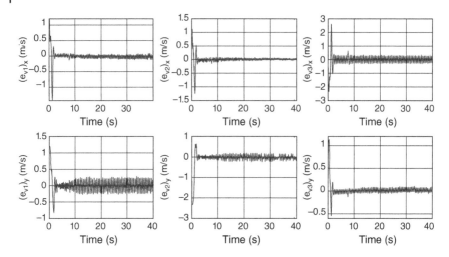

Figure 5.11 Single integrator: formation acquisition. Velocity error $e_v(t)$ for each UGV along x and y directions.

Formation Translation: In this experiment, the UGVs were commanded to move north-east along a bearing of approximately 32° with a constant translational speed of 0.5 m/s while maintaining the inter-vehicle distance of 4 m in the triangular formation. As a result, the desired translational and angular velocities in (2.24) became $v_0 = [0.424, 0.265, 0]$ m/s and $\omega_0 = 0$.

Figure 5.12 shows the position of each UGV as they maneuver on the plane, including snapshots of the actual formation at $t = 6$ s and $t = 20$ s. Figure 5.13 shows the inter-vehicle distance errors converging to a small steady-state value after about 15 s.

The actual translational speed and heading (bearing) angle of each UGV are displayed in Figure 5.14. At first glance, we can see that UGVs 1 and 3 converge to the desired translational speed of 0.5 m/s and the desired bearing angle of 32°, while UGV 2 appears to be moving in the wrong direction. However, based on Figures 5.12 and 5.13, we know that the desired formation maneuvering is achieved. Upon closer inspection, we observe that UGV 2's heading is approximately −148°, which is 180° from the desired heading. Therefore, the vehicle is moving in *reverse* in the direction that promotes the formation maneuvering control objective. This occurs because the formation maneuvering controller is only concerned with the relative position of the vehicles and does not actively control the heading of each vehicle. As a result, the controller may determine that it is beneficial (based on the commanded velocity and the vehicle's current position/heading) to operate in reverse mode.

The DC motor voltage of each vehicle is shown in Figure 5.15. One can observe a steady-state voltage of ±2 V on each of the vehicles. This bias voltage is required to keep the vehicles moving at the desired translational speed

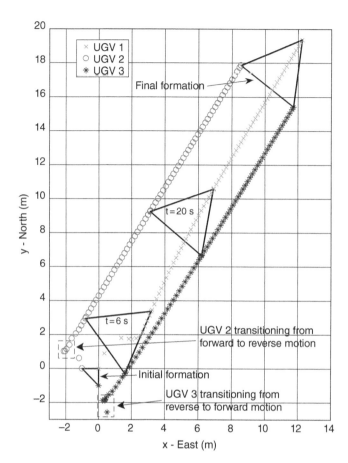

Figure 5.12 Single integrator: formation translation. UGV position trajectories $q_i(t) = [x_i(t), y_i(t)]$, $i = 1, 2, 3$.

of 0.5 m/s. Note that a negative voltage indicates that the vehicle is moving in reverse. Indeed, from Figure 5.12 we can see the moment where UGV 2 transitions from an initial forward motion to the reverse motion, which it maintains until the end of the experiment. On the other hand, UGV 3 does the opposite maneuver: initial reverse motion followed by forward motion. Clearly, the formation translational speed can be increased significantly since each UGV only utilized approximately 2 V/16 V = 12.5% of the battery voltage. The somewhat slow translational speed was specifically chosen to maintain the vehicles operating at low velocities in case of hardware failure, and to prevent the vehicles from overrunning the parking lot area.

The steering angles are displayed in Figure 5.16. Notice that, aside from a brief period during the transient, the steering angles do not saturate as in Figure 5.10

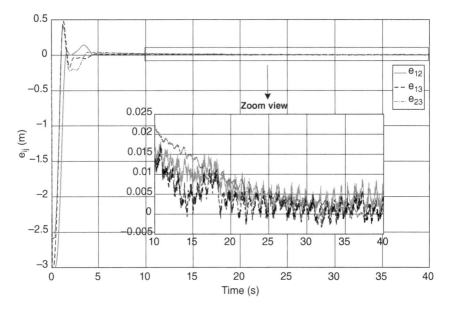

Figure 5.13 Single integrator: formation translation. UGV distance errors $e_{ij}(t), i, j \in \{1, 2, 3\}$.

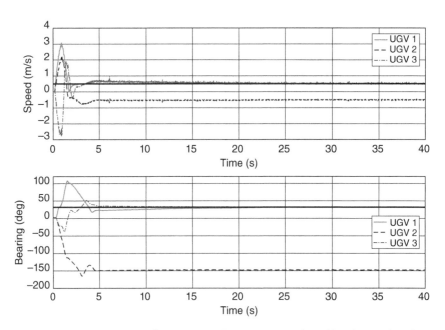

Figure 5.14 Single integrator: formation translation. UGV speeds and heading angles. The thick black lines denote the desired values.

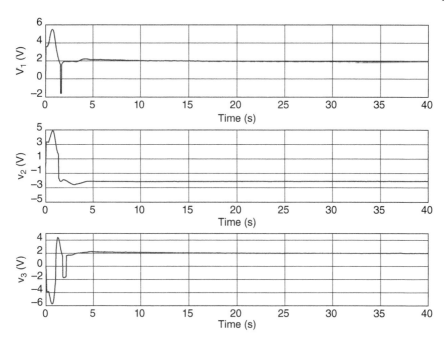

Figure 5.15 Single integrator: formation translation. Drive-wheel DC motor voltages $V_i(t)$, $i = 1, 2, 3$.

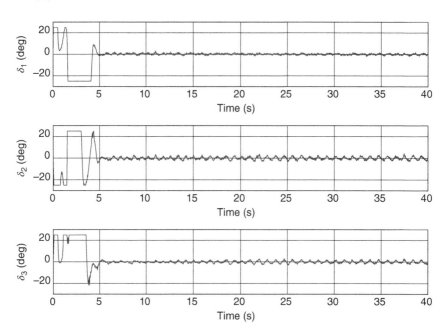

Figure 5.16 Single integrator: formation translation. Steering angle commands $\delta_i(t)$, $i = 1, 2, 3$.

because v_x in (5.10b) never approaches zero during formation maneuvering. The steering angles of each UGV attain a steady-state value of zero since the front-wheel direction needs to eventually align with the X-axis for the vehicle to move in the desired direction.

Formation Translation and Rotation: Here, the formation had to move with the same translational velocity of the previous experiment while rotating about UGV 2 with an angular velocity of $\omega_0 = 20$ deg/s. Figure 5.17 depicts the propagation of the formation over time. Since UGV 2 is the axis of rotation, it merely translates in the north-east direction while UGVs 1 and 3 trace a spiral-like trajectory as they spin around UGV 2. The distance errors in Figure 5.18 indicate that the triangular formation was acquired by 10 s. In Figure 5.19, we can see

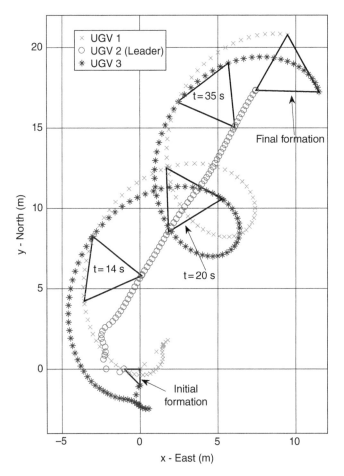

Figure 5.17 Single integrator: formation translation/rotation. UGV position trajectories $q_i(t) = [x_i(t), y_i(t)], i = 1, 2, 3.$

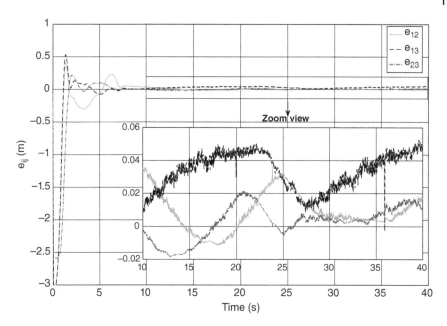

Figure 5.18 Single integrator: formation translation/rotation. UGV distance errors $e_{ij}(t)$, $i,j \in \{1,2,3\}$.

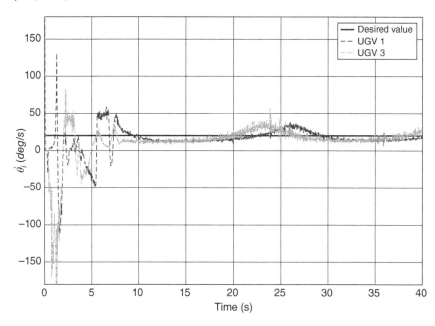

Figure 5.19 Single integrator: formation translation/rotation. UGV bearing rate $\dot{\theta}_i(t)$, $i = 1, 3$.

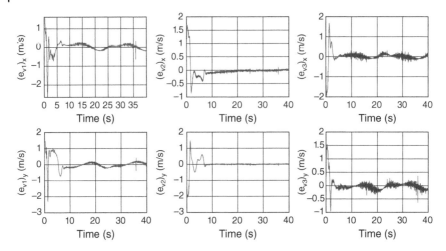

Figure 5.20 Single integrator: formation translation/rotation. Velocity error $e_v(t)$ of each UGV along the x and y directions.

the bearing rate $\dot{\theta}_i(t)$ of UGVs 1 and 3 converging to the desired angular velocity of 20 deg/s. This angle rate was calculated by numerically differentiating the vehicle's measured bearing $\theta_i(t)$ with all discontinuities removed (i.e., when the vehicles' magnetic heading value transitioned from $0°$ to $359°$).

The velocity error of each UGV is shown in Figure 5.20. The smaller error for UGV 2 is expected since its desired velocity is constant (setpoint problem) whereas the desired velocities for UGVs 1 and 3 are time varying (tracking problem). The drive-wheel motor voltages and steering angles are given in Figures 5.21 and 5.22, respectively. From Figure 5.21, we can see that UGV 2 was operating in reverse throughout the experimental trial since its voltage was always negative. Notice the nonzero steady-state values of the steering angles for UGVs 1 and 3 required for producing the rotation about UGV 2. On the other hand, the steering angle of UGV 2 reaches zero at $t \approx 10$ s when the vehicle becomes aligned with the direction of translational velocity vector v_0.

5.4.3 Single Integrator: Target Interception

For this experiment a virtual vehicle was employed as the target since additional UGVs were not available. The target's translational speed was set to a constant value of 0.5 m/s and its heading angle to an oscillatory motion of $32 + 10 \sin(0.2\pi t)$ deg. The same triangular formation from Section 5.4.2 was used. The target velocity estimate $\hat{v}_T$ in (2.45) was initialized to zero and was calculated using the trapezoidal rule of integration. The control gains were set to $k_v = 0.08$ and $k_1 = 1$ in (2.53) and (2.54), and $K_P = 5I_2$ and $K_I = 0.5I_2$ in (5.15). Since only three vehicles were available, UGV 2 was selected as the leader

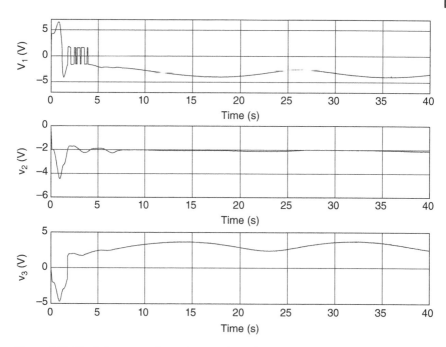

Figure 5.21 Single integrator: formation translation/rotation. Drive-wheel DC motor voltages $V_i(t)$, $i = 1, 2, 3$.

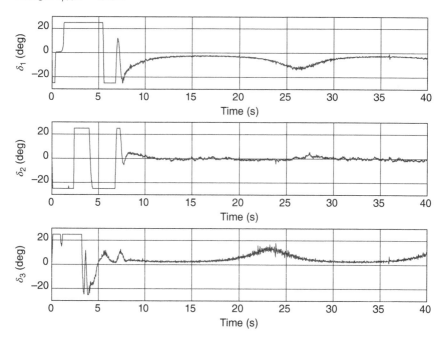

Figure 5.22 Single integrator: formation translation/rotation. Steering angle commands $\delta_i(t)$, $i = 1, 2, 3$.

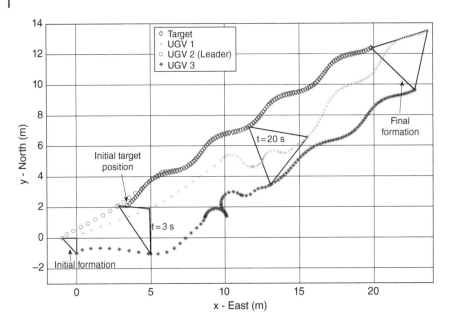

Figure 5.23 Single integrator: target interception. UGV position trajectories $q_i(t) = [x_i(t), y_i(t)]$, $i = 1, 2, 3$ and target position $q_T(t)$. Target was initially positioned at $q_T(0) = (3.4, 2.1)$ m.

of the formation, i.e., the agent responsible for chasing the target. As a result, the target interception error (2.44) was defined as $e_T = q_T - q_2$, and the target was located at the boundary of the formation when intercepted.

Figure 5.23 shows the motions of the UGVs and target over time. The inter-vehicle distance errors and the target interception error defined in (2.44) are given in Figures 5.24 and 5.25, respectively. From these figures, one can see that the target was successfully intercepted after approximately 10 s while the formation was acquired after 20 s.

Figure 5.26 displays the drive-wheel motor voltage for each vehicle. As in the formation maneuvering experiment, the voltages reach a steady-state value of ± 2 V since the formation needs to track the target's translational velocity of 0.5 m/s. The steering angles, shown in Figure 5.27, have an oscillatory motion for $t \geq 20$ s because of the sinusoidal heading movement of the target. That is, the oscillation period is 10 s, which corresponds to the frequency of the target's heading motion (0.1 Hz).

5.4.4 Single Integrator: Dynamic Formation

For the following experiment, the desired formation was set to a time-varying version of the one used in the previous experiments. Specifically, we used an

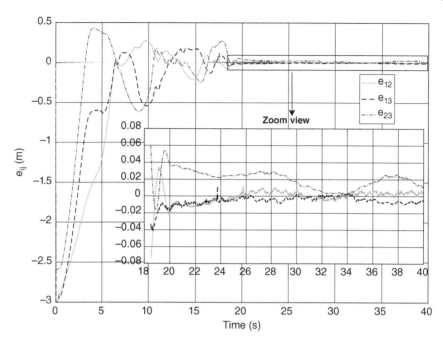

Figure 5.24 Single integrator: target interception. UGV distance errors $e_{ij}(t)$, $i, j \in \{1, 2, 3\}$.

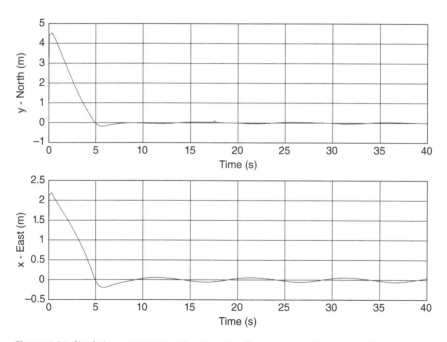

Figure 5.25 Single integrator: target interception. Target interception error $e_T(t)$.

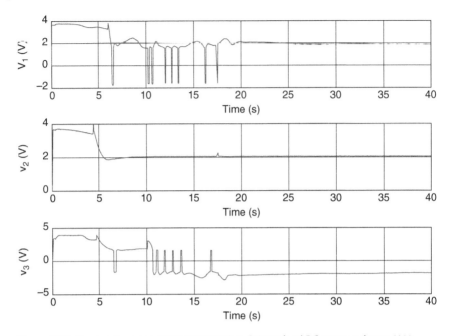

Figure 5.26 Single integrator: target interception. Drive-wheel DC motor voltages $V_i(t)$, $i = 1, 2, 3$.

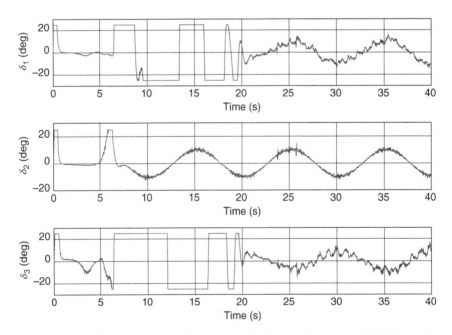

Figure 5.27 Single integrator: target interception. Steering angle commands $\delta_i(t)$, $i = 1, 2, 3$.

equilateral triangle with time-varying inter-vehicle distances specified as

$$d_{ij}(t) = 4 + \sin(0.2\pi t)\,\text{m}. \tag{5.19}$$

In addition, we required the formation to maneuver according to the translation described in the first experiment of Section 5.4.2. The vehicles were again initially located according to (5.17), and the control gains were tuned to $k_v = 2$ in (2.73), and $K_P = 5I_2$ and $K_I = 2I_2$ in (5.15).

The trajectory of each UGV along with snapshots of the formation at instants of minimum and maximum values for (5.19) are given in Figure 5.28. It is interesting to note that UGV 1 strictly followed the desired translation with virtually no oscillations, while UGVs 2 and 3 appear to carry the brunt of the effort to track the sinusoidal dynamic formation. From the distance errors in Figure 5.29,

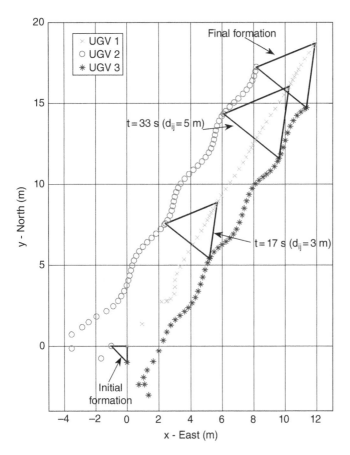

Figure 5.28 Single integrator: dynamic formation. UGV position trajectories $q_i(t) = [x_i(t), y_i(t)], i = 1, 2, 3.$

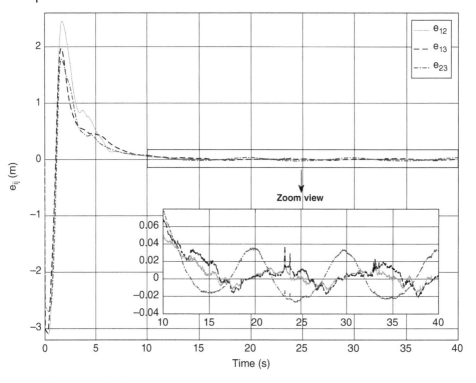

Figure 5.29 Single integrator: dynamic formation. UGV distance errors $e_{ij}(t)$, $i,j \in \{1,2,3\}$.

we see that the desired formation was acquired and tracked after about 15 s with a steady-state error within ± 3.5 cm.

The drive-wheel motor voltages in Figure 5.30 again converge to ± 2 V as needed to move the vehicles with 0.5 m/s speed. In this experiment, the voltages were saturated to 50% of the maximum battery voltage (8 V). Note that UGV 2 moved in reverse for the entire duration of the experimental run. The steering angles are displayed in Figure 5.31. It was mentioned previously that UGV 1 moved in a fairly straight positional track as compared to UGVs 2 and 3. This observation is substantiated by noting that UGV 1's steering angle has much smaller oscillations about $0°$ as compared to UGVs 2 and 3.

5.4.5 Double Integrator: Formation Acquisition

In order to enable the proper comparison between the double integrator (DI)-based formation acquisition controller and the single integrator (SI)-based one, the control gain k_v in (3.9) was set to the same value reported in Section 5.4.1, while we used $k_a = 5$ in (3.8), which is the same value as K_p of (5.15) in the experiment of Section 5.4.1.

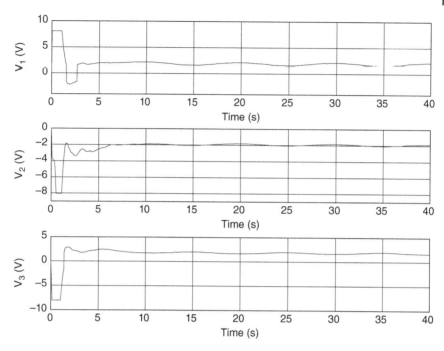

Figure 5.30 Single integrator: dynamic formation. Drive-wheel DC motor voltages $V_i(t)$, $i = 1, 2, 3$.

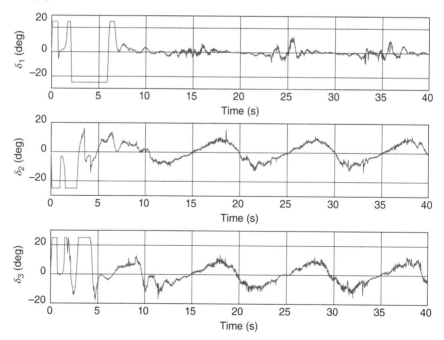

Figure 5.31 Single integrator: dynamic formation. Steering angle commands $\delta_i(t)$, $i = 1, 2, 3$.

The UGV trajectories from their initial configuration to the final configuration are displayed in Figure 5.32, while the inter-vehicle distance errors are shown in Figure 5.34. In Figure 5.33, we replot the trajectories of both the DI- and SI-based controllers to facilitate the comparison. First, one can see that the final formations are isomorphic according to Definition 1.4. Note that the DI-based controller produced a more irregular response than the SI one. This is more clearly seen by the side-by-side comparison of the distance errors in Figure 5.35, where the response of the DI-based controller is noticeably less damped. We believe that this is due to the control gains in (3.8) and (3.9) not being retuned for the DI system, whose closed-loop dynamics have higher order than the SI dynamics. From the inset plots in Figures 5.8 and 5.34, we can see that the steady-state errors of the DI-based control were slightly smaller in their peak-to-peak amplitude than those of the SI-based control. Again, better tuning of DI-based control gains could result in even better steady-state performance.

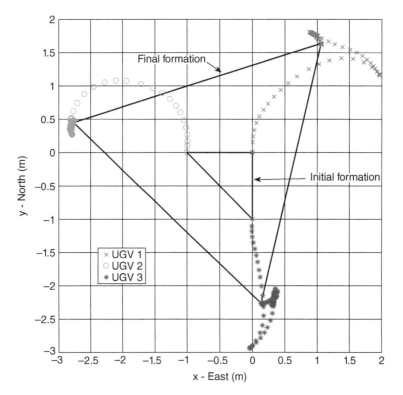

Figure 5.32 Double integrator: formation acquisition. UGV position trajectories $q_i(t) = [x_i(t), y_i(t)], i = 1, 2, 3$.

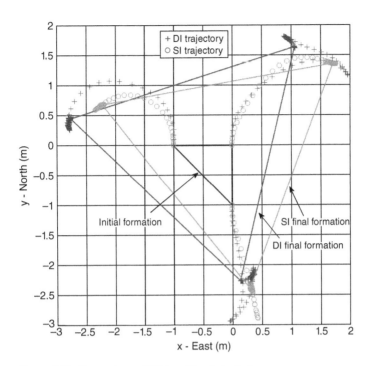

Figure 5.33 Comparison of UGV position trajectories.

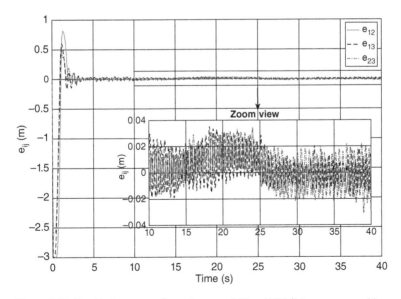

Figure 5.34 Double integrator: formation acquisition. UGV distance errors $e_{ij}(t)$, $i,j \in \{1,2,3\}$.

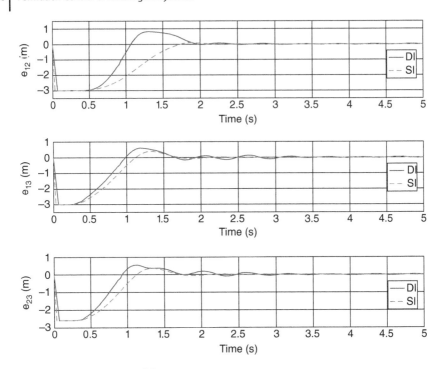

Figure 5.35 Comparison of distance errors.

The motor voltages, which are shown in Figure 5.36, have larger transient magnitudes than those of the SI-based controller (see Figure 5.9); however, the steady-state magnitudes are only slightly larger. This suggests that the DI-based formation acquisition controller required more control energy than its SI counterpart. The steering angles in Figure 5.37 are qualitatively similar to the ones for the SI-based control (see Figure 5.10) in the sense that they operated in saturation for the duration of the experiment due to the UGV longitudinal speed v_x quickly converging to zero.

5.4.6   Double Integrator: Formation Maneuvering

The same translation-only maneuvering experiment described in Section 5.4.2 was run with the DI-based controller. The same control gain values given in Section 5.4.5 were used in (3.17). In order to provide compensation for the uphill grade of the parking lot, the original DI-based control (3.8) was augmented with the following integral feedback term:

$$u = -k_a s - k_I \int s(t)dt + \dot{v}_f - R^{\mathsf{T}}(\tilde{q})z \tag{5.20}$$

where k_I was set to 0.5 for this experiment. This term is akin to the one used in (5.15) for implementation of the SI-based controllers.

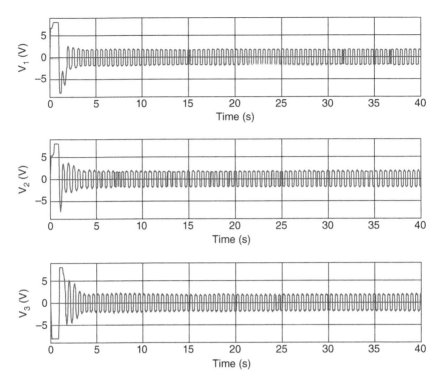

Figure 5.36 Double integrator: formation acquisition. Drive-wheel DC motor voltages $V_i(t)$, $i = 1, 2, 3$.

Figure 5.38 displays the movement of the formation during the 40-s experimental run, and Figure 5.39 compares it to the SI-based control. The trajectories of UGVs 1 and 3 were very similar between the two schemes, while UGV 2 moved initially in opposite directions but then converged to a common trajectory. The distance errors are shown in Figure 5.40, and their comparison with the SI errors are given in Figure 5.41. We again observe the less damped response of the DI errors, which are attributed to the control gains not being fine tuned for the DI dynamics. Indeed, when we calculated the RMS value of the inter-vehicle distance error vector, $e = [e_{12}, e_{13}, e_{23}]$, defined by

$$e_{\text{rms}} = \sqrt{\frac{1}{t_f} \int_0^{t_f} \|e(t)\|^2 \, dt} \tag{5.21}$$

where $t_f = 40$ s is the final experiment time, we obtained $e_{\text{rms}} = 0.660$ for the DI controller as compared to 0.641 when the SI controller was employed.

The velocity error variable s defined in (3.4) is shown in Figure 5.42 for each UGV and converges to zero as predicted by the theory. The motor voltages, displayed in Figure 5.43, have more oscillations and larger magnitude during

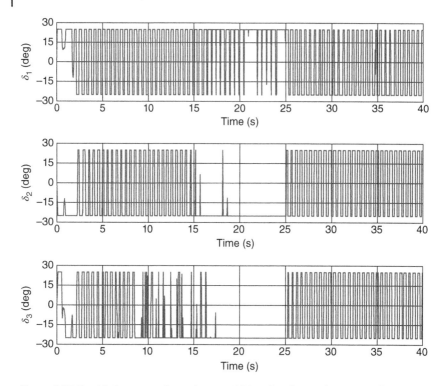

Figure 5.37 Double integrator: formation acquisition. Steering angle commands $\delta_i(t)$, $i = 1, 2, 3$.

the transient period than the SI voltages (see Figure 5.15), but naturally have the same steady-state value of ± 2 V corresponding to the 0.5 m/s translational speed of the maneuver. Note that the oscillations occurred about zero, indicating that the vehicles switched between forward and reverse motions during the first 8–10 s and then settled to one or the other (UGV 1 reverse; UGV 2 reverse; UGV 3 forward) until the end of the experiment. The steering angles in Figure 5.44 are much noisier than the SI ones in Figure 5.16. We believe the DI voltages and steering angles would look similar to their SI counterparts if the control gains were retuned for the DI dynamics.

5.4.7 Double Integrator: Target Interception

The following experiment utilized the same trial conditions as the experiment in Section 5.4.3 with the exception that the target velocity v_T was assumed to be known by the vehicles as per the assumption in Section 3.4. The control law given by (3.19), (3.20), (3.21), and (5.12) was implemented with the addition of the integral feedback term $-k_I \int s(t) dt$ (as was done in (5.20)) and with control

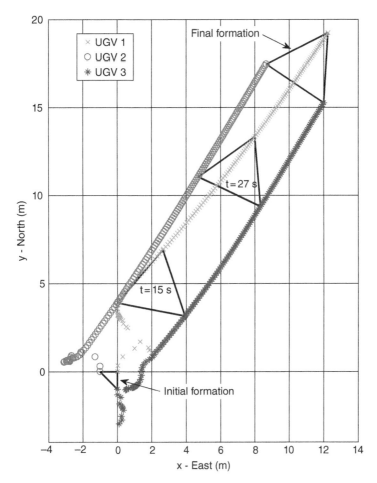

Figure 5.38 Double integrator: formation maneuvering. UGV position trajectories $q_i(t) = [x_i(t), y_i(t)], i = 1, 2, 3$.

gains set to $k_v = 0.08, k_a = 5, k_T = 1.5, k_s = 0$, and $k_I = 0.5$. Notice that the control gain k_s that multiplies the discontinuous term sgn(s) in (3.19) was set to zero in the experiment. During tuning trials, it was observed that larger values of k_s caused the UGVs to jitter; thus, the variable structure-type term was turned off. However, this term would be beneficial if the target motion was more aggressive since it would enable faster reactions from the chasing vehicles.

Figure 5.45 shows the trajectories of the UGVs and virtual target. Figures 5.46 and 5.47 display the inter-vehicle distance errors and target interception error, respectively. These results show that while the desired formation was acquired relatively quickly (by 10 s), the target was only intercepted by UGV 2 after 25 s.

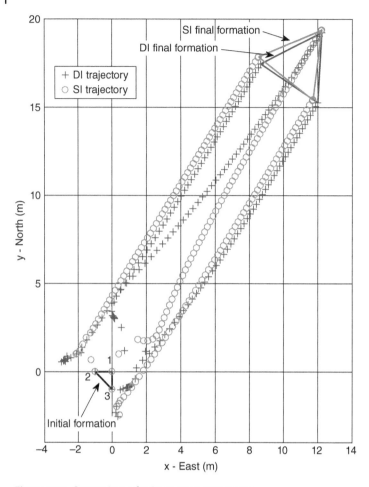

Figure 5.39 Comparison of UGV position trajectories.

The settling time of the target interception error may be explained by the behavior of the velocity error s in Figure 5.48, which also has a settling time of 25 s. This settling time could be shorten by increasing the integral feedback gain k_I from its original value of 0.5 (recall that consistency of control gain values between the various experiments was used to facilitate their comparison). In contrast to the SI target interception error, which exhibited some small steady-state oscillations about zero (see Figure 5.25), the DI target interception error converged to zero with virtually no steady-state offset. This is likely due to the target velocity being known to the DI-based control whereas it was estimated in the SI-based control. As for the inter-vehicle distance errors, the SI and DI controllers yielded similar steady-state values within ±2 cm.

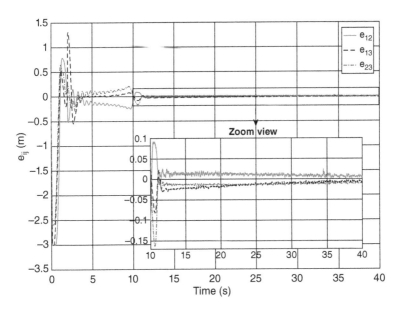

Figure 5.40 Double integrator: formation maneuvering. UGV distance errors $e_{ij}(t)$, $i, j \in \{1, 2, 3\}$.

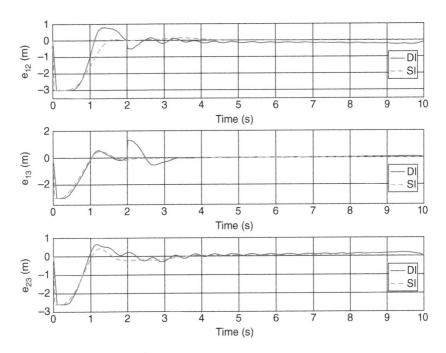

Figure 5.41 Comparison of distance errors.

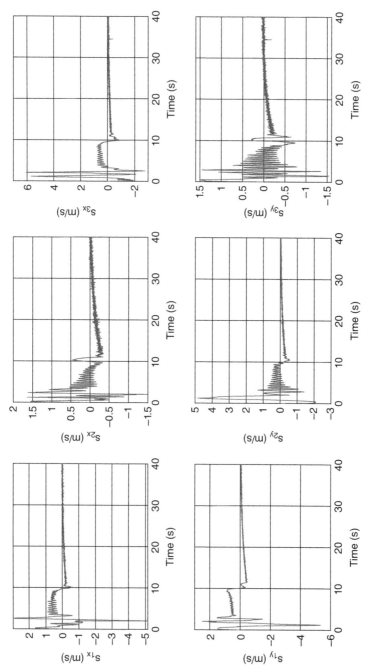

Figure 5.42 Double integrator: formation maneuvering. Velocity error $s(t)$ of each UGV along the x and y directions.

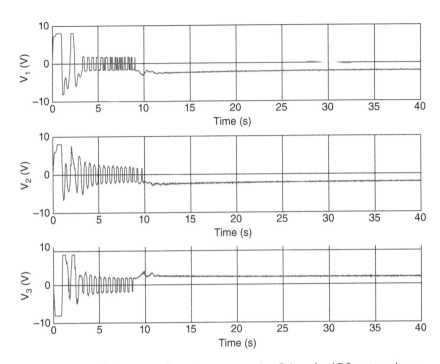

Figure 5.43 Double integrator: formation maneuvering. Drive-wheel DC motor voltages $V_i(t)$, $i = 1, 2, 3$.

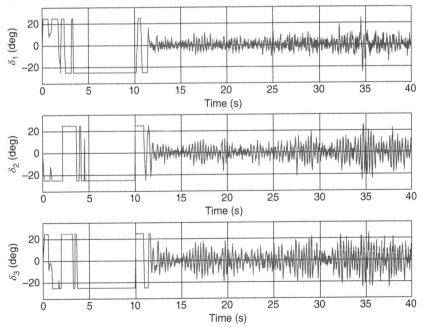

Figure 5.44 Double integrator: formation maneuvering. Steering angle commands $\delta_i(t)$, $i = 1, 2, 3$.

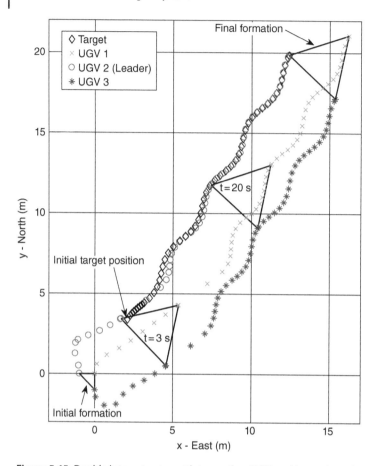

Figure 5.45 Double integrator: target interception. UGV position trajectories $q_i(t) = [x_i(t), y_i(t)]$, $i = 1, 2, 3$ and target position $q_T(t)$. The target was initially positioned at $q_T(0) = (3.4, 2.1)$ m.

The voltages and steering angles are given in Figures 5.49 and 5.50, respectively. In this experiment, the voltages were restricted to 30% of the maximum battery voltage (approximately 4.8 V) to limit the vehicles' velocity to within manageable limits (~ 2 m/s) so that the vehicles stayed within the testing area. As such, the voltages saturated during the first 3 s of the experiment since the transient period is usually the most demanding from the control energy perspective. The DI steering angles are noticeably noisier than the SI ones in Figure 5.27, although one can still distinguish the 10 Hz oscillation caused by the sinusoidal variation of the target heading angle.

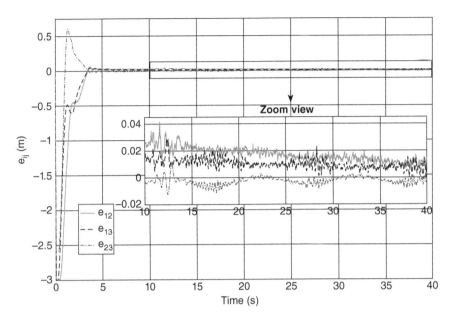

Figure 5.46 Double integrator: target interception. UGV distance errors $e_{ij}(t)$, $i, j \in \{1, 2, 3\}$.

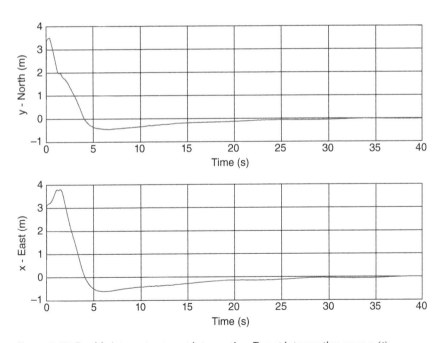

Figure 5.47 Double integrator: target interception. Target interception error $e_T(t)$.

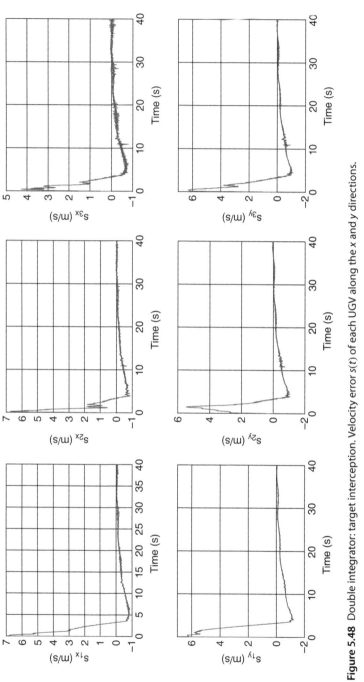

Figure 5.48 Double integrator: target interception. Velocity error $s(t)$ of each UGV along the x and y directions.

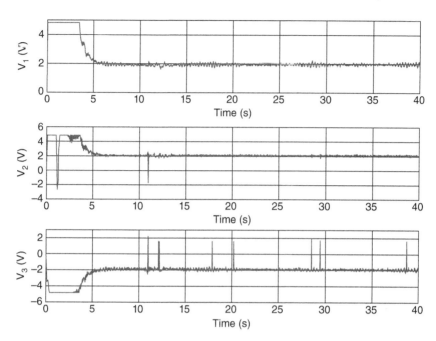

Figure 5.49 Double integrator: target interception. Drive-wheel DC motor voltages $V_i(t)$, $i = 1, 2, 3$.

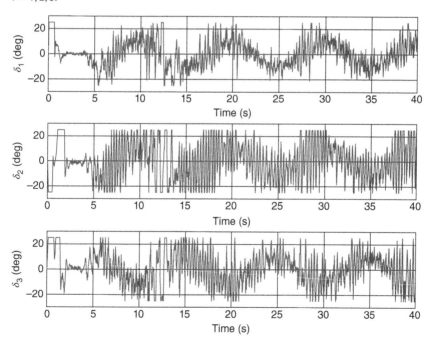

Figure 5.50 Double integrator: target interception. Steering angle commands $\delta_i(t)$, $i = 1, 2, 3$.

5.4.8 Double Integrator: Dynamic Formation

This experiment was conducted using the same formation objectives as Section 5.4.4. The integral feedback term was also included (3.8), as in the previous two DI experiments with control gains set to $k_v = 2$, $k_a = 5$, and $k_I = 0.5$.

The results of the DI-based dynamic formation controller are shown in Figures 5.51 to 5.56 along with comparisons with the SI-based control. Unlike in the SI experiment, UGV 1 did not strictly follow the desired translation and exhibited small oscillations in its heading angle. This is also evidenced by the oscillations of its steering angle seen in Figure 5.56 in comparison to the other UGVs. The settling time and overshoot of the distance errors were smaller for the DI control, although the response was less damped. The DI control also produced smaller steady-state errors, as can be seen by comparing the

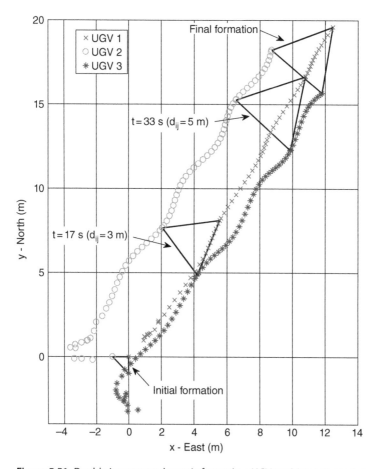

Figure 5.51 Double integrator: dynamic formation. UGV position trajectories $q_i(t) = [x_i(t), y_i(t)]$, $i = 1, 2, 3$.

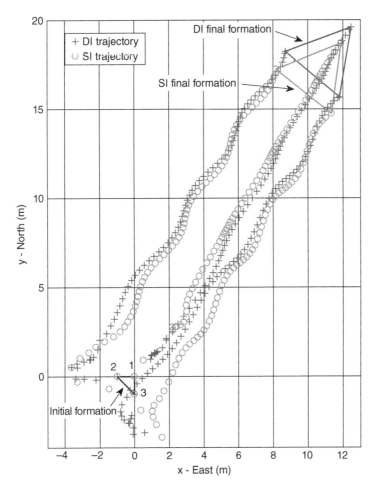

Figure 5.52 Comparison of UGV position trajectories.

insets of Figures 5.53 and 5.29. A comparison of the RMS value of the errors given by (5.21) resulted in $e_{\mathrm{rms}} = 0.742$ for the DI controller and 0.929 for the SI controller. From the drive-wheel motor voltages in Figure 5.55, we can see that the vehicles had more irregular (forward-reverse) motions than when the SI-based control was used (see Figure 5.30 for comparison). Finally, the UGV steering angles of the DI-based control had larger high-frequency variations than those of the SI case.

5.4.9 Holonomic Dynamics: Formation Acquisition

Finally, we tested the adaptive controller presented in Section 4.3.2, which is based on the holonomic dynamics of the UGVs. The hand position of each UGV was set to $L_i = 0.3556$ m, $i = 1, 2, 3$, which is the same value as L_w in (5.9). The

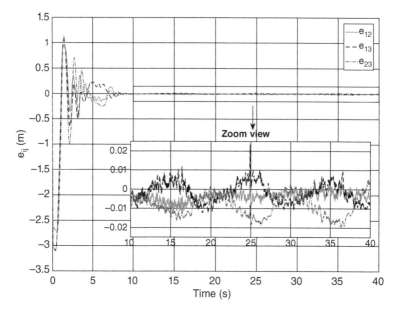

Figure 5.53 Double integrator: dynamic formation. UGV distance errors $e_{ij}(t)$, $i,j \in \{1,2,3\}$.

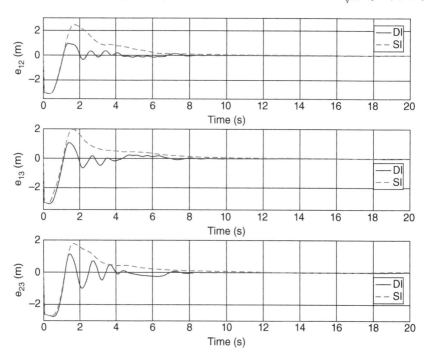

Figure 5.54 Comparison of distance errors.

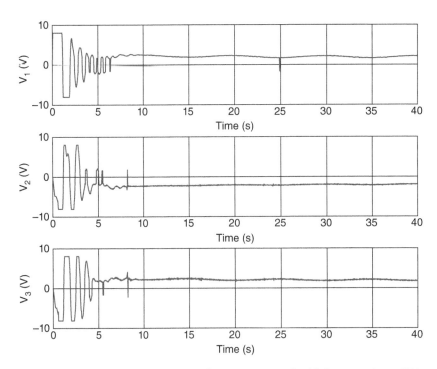

Figure 5.55 Double integrator: dynamic formation. Drive-wheel DC motor voltages $V_i(t)$, $i = 1, 2, 3$.

torque-level control law in (4.30a) was augmented with the integral feedback term $-k_I \int s(t)dt$. The control and adaptation gains in (4.30) and (2.15) were chosen as $k_v = 0.03$, $k_a = 5$, $k_i = 0.5$, and $\Gamma = I_{18 \times 18}$. In addition, the 18 components of the parameter estimate vector $\hat{\phi}(t)$ were initialized to zero.

The motion of the UGVs from the initial formation until the desired formation was acquired is shown in Figure 5.57. From Figure 5.58, we can see that the desired formation was reached within 5 s. The comparison with the SI- and DI-based formation acquisition controllers in Figure 5.59 shows that the response of the adaptive controller was the most damped. In fact, the RMS value of the inter-vehicle distance error given in (5.21) for the three controllers resulted in 0.835 (holonomic), 0.688 (SI), and 0.629 (DI). The higher value of the holonomic-based controller is obviously due to its slower response. However, it is possible that the response can be sped up in a critically damped fashion by additional tuning of the control gains. That is, gain tuning is facilitated in this case by the fact that the adaptive controller—given its certainty equivalence-like structure—is quasi-feedback linearizing the hand dynamics (4.19).

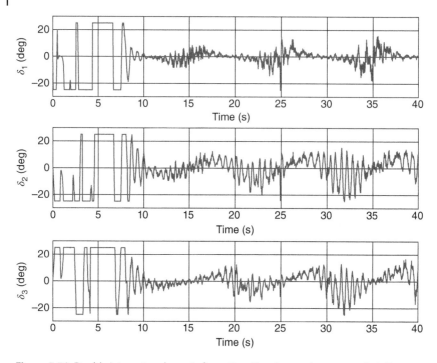

Figure 5.56 Double integrator: dynamic formation. Steering angle commands $\delta_i(t)$, $i = 1, 2, 3$.

For illustration purposes, the six parameter estimates of UGV 1 only are displayed in Figure 5.60. One can observe that the parameter estimates—most notably $[\hat{\phi}_1(t)]_1$ and $[\hat{\phi}_1(t)]_3$—are drifting. The drift is caused by the adaptation law of (4.30b) being primarily driven by the velocity error signal s which does not converge exactly to zero, as shown in Figure 5.61. Since in the pure formation acquisition problem the vehicles operate near zero velocity at steady state, they are subject to noticeable static friction affects. Although the low-level controller attempts to compensate for static friction, the vehicles still tend to jitter about the desired setpoint, as can be seen from the steady-state oscillations in the inset of Figure 5.58. As a result, these oscillations cause the parameter estimates to continue to update. If the experiment were conducted with formation maneuvering, one may not observe the parameter estimate drift. Furthermore, one can always modify the adaptation law with one of the numerous methods for bounding the parameter estimates (e.g., dead zone, σ-modification, e_1-modification, and projection operators [123, 124]).

The drive-wheel voltages and steering angles commands are given in Figures 5.62 and 5.63, respectively. As expected, the voltages are constantly switching after the formation is acquired, albeit at a higher frequency than in

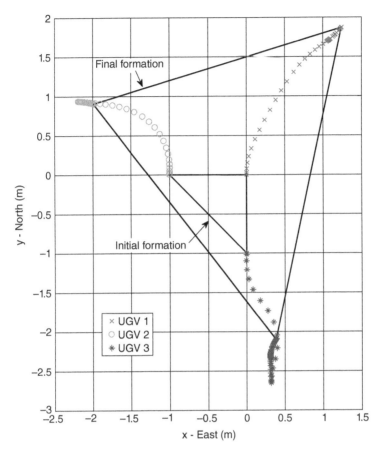

Figure 5.57 Holonomic dynamics: formation acquisition. UGV position trajectories $q_i(t) = [x_i(t), y_i(t)], i = 1, 2, 3.$

the SI and DI formation acquisition experiments. This higher frequency can also be observed in the steering angles. This behavior means that the UGVs experienced a stronger jitter about the desired formation with the adaptive controller.

5.4.10 Summary

The above experimental results provide a baseline for understanding the performance of graph rigidity-based formation controllers. Naturally, the results can be improved by employing better hardware (e.g., more powerful processor, higher resolution optical encoder, wireless router with higher baud rate), more exhaustive control gain tuning, and a more accurate vehicle

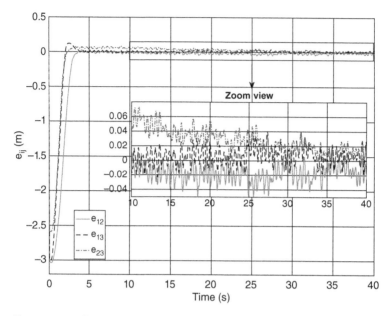

Figure 5.58 Holonomic dynamics: formation acquisition. UGV distance errors $e_{ij}(t)$, $i,j \in \{1,2,3\}$.

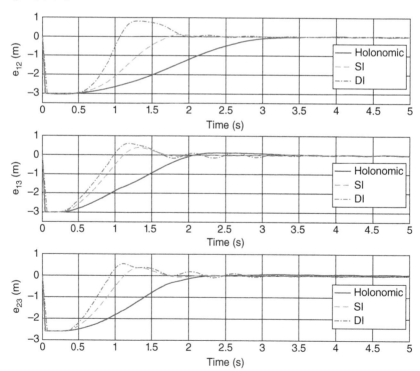

Figure 5.59 Comparison of distance errors.

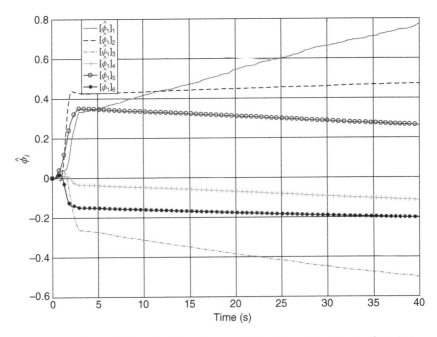

Figure 5.60 Holonomic dynamics: formation acquisition. Parameter estimates for UGV 1, $\hat{\phi}_1(t)$.

model. Nevertheless, a few conclusions can be drawn from the experimental results.

First, the desired formation is easier to obtain when the vehicles are moving (formation maneuvering problem) because static friction is a prominent factor in the stationary case (pure formation acquisition problem). Therefore, adequate static friction compensation is imperative for high-precision applications of UGV formations. Despite the control gains in this experimental study being tuned for the SI-based controllers, the other controllers produced comparable performance. Therefore, there is greater room for improvement in the performance of the DI- and holonomic dynamics-based controllers by better gain tuning. It should be noted that although the control gains have a PID-like interpretation their tuning is not straightforward since the closed-loop system is nonlinear.

Setpoint control problems, such as the formation acquisition problem, can lead to poor transient behavior due to the initial jump of step-like command. One way of improving the transient performance of the formation acquisition controllers is to set the desired distance to the time-varying function, $d_{ij}(t) = \tilde{q}_{ij}(0)\exp(-\alpha t) + \breve{d}_{ij}(1 - \exp(-\alpha t))$ where $\alpha > 0$ is a design parameter and $\breve{d}_{ij}$ is

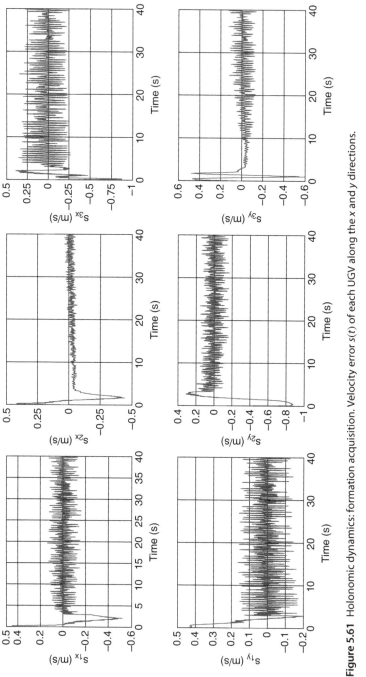

Figure 5.61 Holonomic dynamics: formation acquisition. Velocity error $s(t)$ of each UGV along the x and y directions.

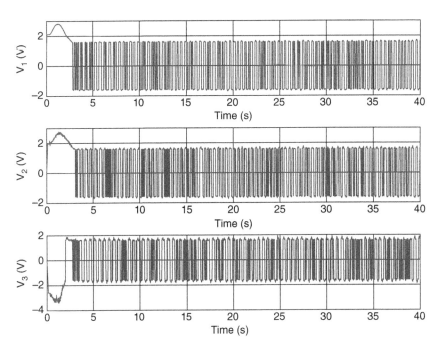

Figure 5.62 Holonomic dynamics: formation acquisition. Drive-wheel DC motor voltages $V_i(t)$, $i = 1, 2, 3$.

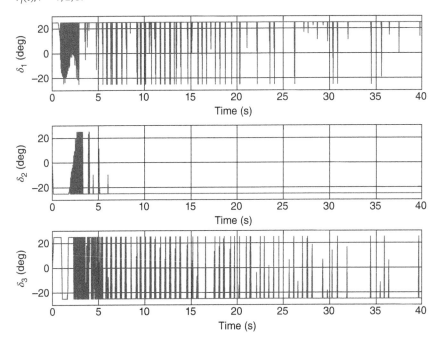

Figure 5.63 Holonomic dynamics: formation acquisition. Steering angle commands $\delta_i(t)$, $i = 1, 2, 3$.

the final desired distance. This function creates a smooth step function with zero initial error.

Finally, independent of what control technique is employed, the importance of a reliable communication system for high-performance formation control cannot be overstated.

Appendix A

Matrix Theory and Linear Algebra

The material in this appendix can be found in, for example, [125–127]. Given a vector $x \in \mathbb{R}^n$, the class of p-norms is defined as

$$\|x\|_p := \left(\sum_{i=1}^{n} |x_i|^p \right)^{1/p}, \quad 1 \le p < \infty.$$

In this book, we primarily use the 1- and 2-norms:

$$\|x\|_1 = \sum_{i=1}^{n} |x_i| \qquad \|x\|_2 = \left(\sum_{i=1}^{n} |x_i|^2 \right)^{1/2} = \sqrt{x^\mathsf{T} x}.$$

If $\|\cdot\|_a$ and $\|\cdot\|_b$ are any two different p-norms, then there exist constants c_1 and c_2 such that

$$c_1 \|x\|_a \le \|x\|_b \le c_2 \|x\|_a, \quad \forall x \in \mathbb{R}^n.$$

For example, it can be found that

$$\|x\|_2 \le \|x\|_1 \le \sqrt{n} \|x\|_2.$$

For any matrix $A \in \mathbb{R}^{n \times n}$, the *eigenvalues* $\lambda_i \in \mathbb{C}$ are the roots of the characteristic equation

$$|\lambda I - A| = 0,$$

and the *eigenvectors* are the nonzero vectors $v_i \in \mathbb{R}^n$ satisfying the equation

$$A v_i = \lambda_i v_i.$$

For an $n \times n$ matrix, there are always exactly n eigenvalues, either complex or real.

Since

$$|A| = \prod_{i=1}^{n} \lambda_i,$$

a matrix is nonsingular if and only if all its eigenvalues are nonzero.

Formation Control of Multi-Agent Systems: A Graph Rigidity Approach, First Edition.
Marcio de Queiroz, Xiaoyu Cai, and Matthew Feemster.
© 2019 John Wiley & Sons Ltd. Published 2019 by John Wiley & Sons Ltd.
Companion website: www.wiley.com/go/dequeiroz/formation_control

A matrix $A = [a_{ij}] \in \mathbb{R}^{n \times n}$ is *symmetric* if and only if $A^\top = A$, i.e., $[a_{ij}] = [a_{ji}]$ for $i \neq j$, and is *skew-symmetric* if and only if $A^\top = -A$, i.e., $[a_{ij}] = -[a_{ji}]$ for $i \neq j$ and $[a_{ii}] = 0$. A skew-symmetric matrix A has the property that

$$x^\top A x = 0, \ \forall x \in \mathbb{R}^n$$

Given $a, b \in \mathbb{R}^3$, we have that

$$a \times b = S(a)b$$

where $S(a)$ is the skew-symmetric matrix

$$S(a) = \begin{bmatrix} 0 & -a_3 & a_2 \\ a_3 & 0 & -a_1 \\ -a_2 & a_1 & 0 \end{bmatrix}.$$

The following result is known as the *Rayleigh–Ritz Theorem*.

Theorem A.1 *[126]* If $A \in \mathbb{R}^{n \times n}$ is symmetric, then

$$\lambda_{\min}(A)x^\top x \leq x^\top A x \leq \lambda_{\max}(A)x^\top x, \forall x \in \mathbb{R}^n$$

where

$$\lambda_{\min}(A) = \min\{\lambda_i\} = \min_{x \neq 0} \frac{x^\top A x}{x^\top x} \quad \text{and} \quad \lambda_{\max}(A) = \max\{\lambda_i\} = \min_{x \neq 0} \frac{x^\top A x}{x^\top x}.$$

The *Kronecker product* of matrices $A = [a_{ij}] \in \mathbb{R}^{m \times n}$ and $B = [b_{ij}] \in \mathbb{R}^{p \times q}$ is denoted by $A \otimes B$ and defined as the following $mp \times nq$ matrix

$$A \otimes B := \begin{bmatrix} a_{11}B & \cdots & a_{1n}B \\ \vdots & \vdots & \vdots \\ a_{m1}B & \cdots & a_{mn}B \end{bmatrix}.$$

The *direct sum* of matrices $A_i \in \mathbb{R}^{m_i \times n_i}$, $i = 1, \dots, p$ is defined to be the $(m_1 + \cdots + m_p) \times (n_1 + \cdots + n_p)$ matrix

$$A_1 \oplus A_2 \oplus \cdots \oplus A_p = \begin{bmatrix} A_1 & 0 & \cdots & 0 \\ 0 & A_2 & \cdots & 0 \\ \vdots & \vdots & \ddots & \vdots \\ 0 & \cdots & 0 & A_p \end{bmatrix}.$$

The *null space* of $A \in \mathbb{R}^{m \times n}$ is the set of solutions to the homogeneous equation $Ax = 0$, i.e.,

$$\text{Null}(A) := \{x \in \mathbb{R}^n \mid Ax = 0\}$$

where 0 here is the $n \times 1$ zero vector. Solutions $x \neq 0$ for the homogeneous equation are called *nontrivial* solutions. The *range* of $A \in \mathbb{R}^{m \times n}$ is defined by the set

$$\text{Range } (A) := \{y \in \mathbb{R}^m \mid y = Ax\}.$$

The following relation holds between these two subspaces

$$n = \dim(\text{Null}(A)) + \dim(\text{Range}(A)).$$

The *rank* of matrix $A \in \mathbb{R}^{m \times n}$ is the largest number of linearly independent rows (or columns) of A. Therefore, we have that

$$\text{rank}(A) \leq \min\{m, n\}.$$

We also know that

$$\text{Range}(A) = \text{rank}(A)$$

and therefore,

$$\text{rank}(A) = n - \dim(\text{Null}(A)).$$

A useful rank equality is

$$\text{rank}(A) = \text{rank}(A^\top) = \text{rank}(AA^\top) = \text{rank}(A^\top A).$$

A pseudoinverse of a matrix $A \in \mathbb{R}^{m \times n}$ is a generalization of the inverse matrix. The most widely known pseudoinverse is the *Moore–Penrose pseudoinverse*. The pseudoinverse $A^+ \in \mathbb{R}^{n \times m}$ is a matrix satisfying the following conditions

$$AA^+A = A$$
$$A^+A\,A^+ = A^+$$
$$(A^+A)^\top = A^+A$$
$$(AA^+)^\top = AA^+.$$

The pseudoinverse exists for any matrix A, but when A has full rank, A^+ has a simple algebraic formula. Specifically, if A has linearly independent columns, then the matrix $A^\top A$ is invertible and

$$A^+ = (A^\top A)^{-1}A^\top.$$

In this case, the pseudoinverse is called a left inverse since $A^+A = I_n$. If A has linearly independent rows ($AA^\top$ is invertible), then

$$A^+ = A^\top(AA^\top)^{-1},$$

which is called a right inverse since $AA^+ = I_m$.

Appendix B

Functions and Signals

The material in this appendix can be found in [8, 128]. A function $f : \mathbb{R}^n \to \mathbb{R}^m$ is said to be *continuously differentiable* ($f \in C^1$) at a point x_0 if the partial derivatives $\partial f_i \backslash \partial x_j$ exist and are continuous at x_0 for $1 \le i \le m$ and $1 \le j \le n$. The function is continuously differentiable on a set $S \subseteq \mathbb{R}^n$ if it is continuously differentiable at every point of S. A function belongs to C^p, $p \ge 1$ on S if f_i has continuous partial derivatives up to order p on S. If $f \in C^\infty$, we say the function is *sufficiently smooth*.

For any piecewise continuous signal $x : \mathbb{R}_{\ge 0} \to \mathbb{R}^n$,

$$\|x\|_{\mathcal{L}_\infty} := \sup_{t \ge 0} \|x(t)\| \quad \text{and} \quad \|x\|_{\mathcal{L}_2} := \sqrt{\int_0^\infty \|x(t)\|^2 dt}.$$

If $\|x\|_{\mathcal{L}_\infty} < \infty$ (the signal is bounded for all time), we say that $x(t) \in \mathcal{L}_\infty$. Likewise, if $\|x\|_{\mathcal{L}_2} < \infty$ (the signal is square-integrable), we say that $x(t) \in \mathcal{L}_2$. Note that any p-norm $\|\cdot\|$ may be used in the above definitions; however, it is common to define the $\mathcal{L}_2$ space with the Euclidean norm.

Given a differentiable signal $x : \mathbb{R}_{\ge 0} \to \mathbb{R}$, if its time derivative satisfies the relationship

$$|\dot{x}(t)| \le \beta_0 \exp(-\beta_1 t), \quad \forall t \ge 0 \tag{B.1}$$

for $\beta_0, \beta_1 > 0$, then $x(t) \in \mathcal{L}_\infty$.

A continuous function $\alpha : [0, a] \to \mathbb{R}_{\ge 0}$ is said to belong to class $\mathcal{K}$ if it is strictly increasing and $\alpha(0) = 0$. A continuous function $\beta : [0, a) \times \mathbb{R}_{\ge 0} \to \mathbb{R}_{\ge 0}$ is said to belong to class $\mathcal{KL}$ if, for each fixed s, the mapping $\beta(r, s)$ belongs to class $\mathcal{K}$ with respect to r and, for each fixed r, the mapping $\beta(r, s)$ is decreasing with respect to s and $\beta(r, s) \to 0$ as $s \to \infty$.

Formation Control of Multi-Agent Systems: A Graph Rigidity Approach, First Edition.
Marcio de Queiroz, Xiaoyu Cai, and Matthew Feemster.
© 2019 John Wiley & Sons Ltd. Published 2019 by John Wiley & Sons Ltd.
Companion website: www.wiley.com/go/dequeiroz/formation_control

Appendix C

Systems Theory

C.1 Linear Systems

The following material can be found in [129, 130]. Given a real function of time $f(t)$ satisfying the condition

$$\int_{0^-}^{\infty} |f(t)|e^{-\sigma t}dt < \infty$$

for some finite real σ, its *Laplace transform* is defined as

$$\mathcal{L}[f(t)] = F(s) := \int_{0^-}^{\infty} f(t)e^{-st}dt,$$

where $s = \sigma + j\omega$ (complex number) is called the Laplace variable.

Two important properties of the Laplace transform are

$$\mathcal{L}\left[\frac{df(t)}{dt}\right] = sF(s) - f(0^-)$$

$$\mathcal{L}\left[\int_{0^-}^{t} f(\tau)d\tau\right] = \frac{F(s)}{s}.$$

Roughly speaking, the above properties indicate that multiplication by s in the Laplace domain is equivalent to the differential operator in the time domain ($s \leftrightarrow d/dt$). Likewise, division by s in the Laplace domain is equivalent to the integral operator in the time domain ($1/s \leftrightarrow \int dt$).

Consider the single-input/single-output (SISO), linear time-invariant (LTI) system

$$\frac{d^n y}{dt^n} + a_{n-1}\frac{d^{n-1}y}{dt^{n-1}} + \ldots + a_1\frac{dy}{dt} + a_0 y = b_m\frac{d^m u}{dt^m} + \ldots + b_1\frac{du}{dt} + b_0 u,$$

with initial conditions $y(0), \ldots, d^{n-1}y(0)/dt^{n-1}$, where a_i and b_i are real constants, $y(t)$ is the scalar output, and $u(t)$ is the scalar input. The *transfer function*

Formation Control of Multi-Agent Systems: A Graph Rigidity Approach, First Edition.
Marcio de Queiroz, Xiaoyu Cai, and Matthew Feemster.
© 2019 John Wiley & Sons Ltd. Published 2019 by John Wiley & Sons Ltd.
Companion website: www.wiley.com/go/dequeiroz/formation_control

of the system is defined as the ratio of the Laplace transform of the output over the Laplace transform of the input, with all initial conditions assumed to be zero. That is,

$$G(s) = \frac{Y(s)}{U(s)} = \frac{b_m s^m + \ldots + b_1 s + b_0}{s^n + a_{n-1} s^{n-1} + \ldots + a_1 s + a_0}.$$

In general, if an LTI system has p inputs and q outputs, the transfer function between the jth input and the ith output is defined as

$$G_{ij}(s) = \frac{Y_i(s)}{U_j(s)}$$

with $U_k(s) = 0$, $k = 1, \ldots, p$, $k \neq j$ (i.e., all inputs other than the jth are set to zero). In matrix-vector form, we then have that

$$Y(s) = G(s)U(s)$$

where $Y(s) = [Y_1(s), \ldots, Y_q(s)]$, $U(s) = [U_1(s), \ldots, U_p(s)]$, and

$$G(s) = \begin{bmatrix} G_{11}(s) & \cdots & G_{1p}(s) \\ \vdots & \vdots & \vdots \\ G_{q1}(s) & \cdots & G_{qp}(s) \end{bmatrix}$$

is the $q \times p$ transfer function matrix.

The following theorem is a valuable result for input-output stability.

Theorem C.1 *[129]* Consider the SISO LTI system

$$\dot{x} = Ax + Bu$$
$$y = Cx$$

where $A \in \mathbb{R}^{n \times n}$ is a Hurwitz matrix. Then, the following results hold:

- If $u(t) \in \mathcal{L}_2$, then $y(t) \in \mathcal{L}_2 \cap \mathcal{L}_\infty$, $\dot{y}(t) \in \mathcal{L}_2$, $y(t)$ is continuous, and $y(t) \to 0$ as $t \to \infty$.
- If $u(t) \in \mathcal{L}_\infty$, then $y(t) \in \mathcal{L}_\infty$, $\dot{y}(t) \in \mathcal{L}_\infty$, and $y(t)$ is uniformly continuous. If, in addition, $u(t) \to 0$ as $t \to \infty$, then $y(t) \to 0$ as $t \to \infty$.

C.2   Nonlinear Systems

The following material can be found in [8, 9, 131]. Consider the nonautonomous (time-varying) system

$$\dot{x} = f(x, t), \quad x(t_0) = x_0 \tag{C.1}$$

where $f : D \times \mathbb{R}_{\geq 0} \to \mathbb{R}^n$ is locally Lipschitz in x and piecewise continuous in t on $D \times \mathbb{R}_{\geq 0}$. Since f is in general a nonlinear function of t and x, we seek to *qualify* the stability properties of (C.1).

The first step in this analysis is to determine the equilibrium points of the system. A point $x_e(t) \in D$ is an *equilibrium point* of (C.1) at $t = t_0 \geq 0$ if it has the property that whenever the system state starts at the equilibrium, it remains at the equilibrium for all $t > t_0$. Mathematically, this means that equilibrium points can be found by solving the algebraic equation

$$f(x, t) = 0.$$

A nonlinear system may have a unique equilibrium point, a finite number of equilibrium points, or an infinite number of equilibrium points. In the case of multiple equilibrium points, each one could have a different stability property. The issue of stability deals with the behavior of the solutions of (C.1) for initial conditions away from an equilibrium point (i.e., $x_0 \neq x_e(t_0)$). That is, does an equilibrium point attract the solution, repel the solution, or neither (e.g., a periodic solution). It is standard practice in stability analysis to shift a nonzero equilibrium point of interest to the origin through the variable transformation

$$z(t) = x(t) - x_e(t),$$

such that (C.1) with equilibrium point $x_e \neq 0$ is equivalent to $\dot{z} = g(z, t)$ with equilibrium point $z_e = 0$. Henceforth, we will assume 0 is an equilibrium point of (C.1).

Let the set

$$B(\bar{x}, r) = \{x \in \mathbb{R}^n : \|x - \bar{x}\| < r\} \tag{C.2a}$$

represent the "ball" of radius r centered at $\bar{x}$. Stability properties of the equilibrium point of (C.1) are said to hold:

- *locally* if they are true for all $x_0 \in B(0, r)$
- *globally* if they are true for all $x_0 \in \mathbb{R}^n$[1]
- *semi-globally* if they are true for all $x_0 \in B(0, r)$ with arbitrary r
- *uniformly* if they are true for all initial times $t_0 \geq 0$.

The stability properties of the controllers developed in this book are true only locally.

The equilibrium point $x = 0$ of (C.1) is said to be:

- *uniformly stable* if, given any $\varepsilon > 0$, there exists $\delta(\varepsilon) > 0$ (independent of t_0) such that

$$\|x(t_0)\| \in B(0, \delta) \Rightarrow \|x(t)\| \in B(0, \varepsilon), \quad \forall t \geq t_0$$

- *unstable* if it is not stable

---

1 Only systems with a *unique* equilibrium point can have global stability properties.

- *uniformly convergent* if there exists $\delta > 0$ such that

$$\|x(t_0)\| \in B(0, \delta) \Rightarrow x(t) \to 0 \text{ as } t \to \infty$$

- *uniformly asymptotically stable* if it is both uniformly stable and uniformly convergent
- *exponentially stable* if there exist $\delta, c_1, c_2 > 0$ such that

$$\|x(t_0)\| \in B(0, \delta) \Rightarrow \|x(t)\| \leq c_1 \|x(t_0)\| \exp(-c_2(t - t_0)), \quad \forall t \geq t_0. \tag{C.3}$$

Autonomous (time-invariant) systems are a special case of (C.1) where the right-hand side of the differential equation is not explicitly dependent on time, i.e., $\dot{x} = f(x)$. Therefore, equilibrium points of autonomous systems are always constant (time independent). Since the solution of an autonomous system depends only on $t - t_0$, the stability properties of its equilibrium points are always uniform and t_0 can be taken as zero without loss of generality. For similar reasons, the qualifier "uniform" is not necessary when referring to the exponential stability of nonautonomous systems (notice that $t - t_0$ appears in (C.3)).

C.3 Lyapunov Stability

The following material can be found in [8, 9, 131, 132]. Lyapunov theory enables one to qualitatively assess the stability properties of an equilibrium point of interest without having to explicitly solve the nonlinear differential equation (C.1). Specifically, the so-called *Lyapunov's second (or direct) method* is based on the following, fundamental physical observation [132]: If a system's total energy is continuously dissipated, then the system must eventually settle down to an equilibrium point. That is, equilibrium points are zero-energy points. Since energy is a scalar quantity, we can study the stability of a system by examining the time variation of a single scalar function that captures the total energy of the system. In the case of mechanical systems (which multi-agent systems fall under), this function should be related to the potential energy (position dependent) and kinetic energy (velocity dependent). This energy-like function is known as the *Lyapunov function candidate*.

The notion of positive definite functions (and its variants) plays an important role in Lyapunov's second method. A function $V : D \to \mathbb{R}$ where $0 \in D \subseteq \mathbb{R}^n$ is said to be:

- *positive definite* in D if $V(x) > 0$ for all $x \in D - \{0\}$ and $V(0) = 0$
- *positive semi-definite* in D if $V(x) \geq 0$ for all $x \in D - \{0\}$ and $V(0) = 0$

- *negative definite* in D if $-V(x)$ is positive definite
- *negative semi-definite* in D if $-V(x)$ is positive semi-definite.

The simplest and most important type of positive definite function is the so-called *quadratic function*:

$$V(x) = x^{\top} A x$$

where $A \in \mathbb{R}^{n \times n}$ is symmetric. In the case of quadratic functions, checking the sign definiteness of $V(x)$ is quite easy. Specifically, $V(x)$ (or matrix A) is:

- positive definite if all eigenvalues of A are positive
- positive semi-definite if all eigenvalues of A are nonnegative
- negative definite if all eigenvalues of A are negative
- negative semi-definite if all eigenvalues of A are nonpositive
- indefinite if some eigenvalues of A are positive and some are negative.

Lemma C.1 *[126]* Suppose that a symmetric matrix A is partitioned as follows

$$A = \begin{bmatrix} A_1 & A_2 \\ A_2^{\top} & A_3 \end{bmatrix}$$

where A_1 and A_3 are square. Then, A is positive definite if and only if both A_1 and $A_3 - A_2^{\top} A_1^{-1} A_2$ are positive definite.

We are now ready to state some Lyapunov stability results. These results are based on the simple mathematical fact that if a scalar function is both bounded from below and decreasing, the function has a limit as time approaches infinity. In the following, we assume $x = 0$ is an equilibrium point for (C.1) and $D \subseteq \mathbb{R}^n$ is a set containing $x = 0$.

Theorem C.2 *[8]* Let $V : D \times \mathbb{R}_{\geq 0} \to \mathbb{R}$ be a continuously differentiable function such that

$$U_1(x) \leq V(x, t) \leq U_2(x)$$

$$\dot{V} = \frac{\partial V}{\partial t} + \frac{\partial V}{\partial x} f(x, t) \leq -U_3(x)$$

for all $t \geq 0$ and for all $x \in D$, where $U_i(x)$, $i = 1, 2, 3$ are continuous positive definite functions on D. Then, $x_e = 0$ is uniformly asymptotically stable.

Corollary C.1 *[8]* If $U_i(x) = c_i \|x\|^p$, $i = 1, 2, 3$ where $c_i, p > 0$ in Theorem C.2, then $x_e = 0$ is exponentially stable.

The following theorem is a corollary to Barbalat's Lemma.

Theorem C.3 [9] Consider the function $f : \mathbb{R}_{\geq 0} \to \mathbb{R}$. If $f(t) \in \mathcal{L}_\infty$, $\dot{f}(t) \in \mathcal{L}_\infty$, and $f(t) \in \mathcal{L}_2$, then

$$f(t) \to 0 \text{ as } t \to \infty.$$

C.4 Input-to-State Stability

The following material can be found in [133, 134]. Input-to-state stability bridges the gap between the notions of Lyapunov stability and input–output stability by quantifying the effects of both initial conditions and external (control or disturbance) inputs on the system state.

Consider the system

$$\dot{x} = f(x, u), \quad x(0) = x_0 \tag{C.4}$$

where $f : \mathbb{R}^n \times \mathbb{R}^m \to \mathbb{R}^n$ is locally Lipschitz in x and u. The input $u(t)$ is a piecewise continuous, bounded function for all $t \geq 0$. System (C.4) is said to be *input-to-state stable* if there exist a class $\mathcal{KL}$ function β and a class $\mathcal{K}$ function α such that, for any x_0 and any $u(t) \in \mathcal{L}_\infty$, the solution $x(t)$ exists for all $t \geq 0$ and satisfies

$$\|x(t)\| \leq \beta(\|x_0\|, t) + \alpha(\sup_{0 \leq \tau \leq t} \|u(\tau)\|). \tag{C.5}$$

The above inequality has several implications.

- For any bounded input, the state is bounded.
- As $t \to \infty$, the state is ultimately bounded by function α.
- If $u(t) \to 0$ as $t \to \infty$, so does $x(t)$.

Theorem C.4 [133] Consider that $f(x, u)$ in (C.4) is locally Lipschitz in (x, u) in some neighborhood of $(x = 0, u = 0)$. Then, the system is locally input-to-state stable if and only if the unforced system $\dot{x} = f(x, 0)$ has a locally asymptotically stable equilibrium point at the origin.

Theorem C.5 [134] Consider the interconnected system

$$\begin{aligned} \Sigma_1 &: \quad \dot{x} = f(x, y) \\ \Sigma_2 &: \quad \dot{y} = g(y). \end{aligned} \tag{C.6}$$

If subsystem Σ_1 with input y is locally input-to-state stable and $y = 0$ is a locally asymptotically stable equilibrium point of subsystem Σ_2, then $[x, y] = 0$ is a locally asymptotically stable equilibrium point of the interconnected system.

C.5   Nonsmooth Systems

The following material can be found in [38, 42, 135, 136]. Consider system

$$\dot{x} = f(x,t), \quad x(t_0) = x_0 \tag{C.7}$$

where $f : D \times \mathbb{R}_{\geq 0} \to \mathbb{R}^n$ is *discontinuous* in x and piecewise continuous in t on $D \times \mathbb{R}_{\geq 0}$. Unfortunately, classical analysis methods are not applicable to differential equations with discontinuous right-hand side (a.k.a. nonsmooth systems) since they require $f(x,t)$ to be at least Lipschitz in x. For such differential equations, even the notion of existence of solutions has to be redefined. A key contribution to this problem was made by Filippov, who developed a solution concept that only requires $f(x,t)$ to be Lebesgue measurable with respect to x and t. This solution is usually called a generalized or Filippov solution. The discontinuities that appear in $f(x,t)$ in this book are of the type $sgn(x)$ which admit a Filippov solution.

A Filippov solution is found by embedding $f(x,t)$ into a set-valued map $\mathcal{F}(x,t)$, and then investigating the existence of a solution to the so-called *differential inclusion*

$$\dot{x} \in \mathcal{F}(x,t), \quad x(t_0) = x_0.$$

A natural choice for this set-valued map is the closed convex hull of $f(x,t)$. If for any t, $0 \in \mathcal{F}(x,t)$, then $x = 0$ is an equilibrium point of (C.7).

Definition C.1 *(Filippov Solution)* Let $f(x,t)$ in (C.7) be Lebesgue measurable and essentially locally bounded,[2] uniformly in t and D be an open and connected set. A function $x : \mathbb{R}_{\geq 0} \to \mathbb{R}^n$ is called a Filippov solution of (C.7) on $[0, \infty)$ if $x(t)$ is absolutely continuous and for almost all $t \in \mathbb{R}_{\geq 0}$

$$\dot{x} \in K[f](x,t),$$

where $K[f](x,t)$ is an upper semi-continuous, nonempty, compact, convex-valued map on D defined as

$$K[f](x,t) := \bigcap_{\delta > 0} \bigcap_{\mu N = 0} \overline{co}(B(x,\delta)\backslash N, t),$$

where $\bigcap_{\mu N = 0}$ denotes the intersection over all sets N of Lebesgue measure zero, $\overline{co}$ is the convex closure, and B was defined in (C.2a).

In order to conduct a Lyapunov analysis of equilibria of a differential inclusion, we can invoke the following result from [42].

---

2 This means the function is bounded on a bounded neighborhood of every point excluding sets of measure zero.

Theorem C.6   [42] If $0 \in K[f](0, t)$ in a region $Q \supset B(0, \delta) \times [t_0, \infty)$ and $V :$ $D \times \mathbb{R}_{\geq 0} \to \mathbb{R}$ is a regular function satisfying $V(0, t) = 0$,

$$\alpha_1(\|x\|) \leq V(x, t) \leq \alpha_2(\|x\|) \quad \forall x \neq 0$$

and

$$\dot{V} \stackrel{a.e.}{\in} \bigcap_{\xi \in \partial V(x,t)} \xi^T \begin{bmatrix} K[f](x, t) \\ 1 \end{bmatrix} \leq -\alpha_3(\|x\|)$$

in Q where $\alpha_i(\cdot)$, $i = 1, 2, 3$ are class $\mathcal{K}$ functions, then $x = 0$ is a uniformly asymptotically stable equilibrium point of (C.7).

C.6   Integrator Backstepping

The following material can be found in [8, 9, 131]. *Integrator backstepping* is a recursive control design methodology for systems in so-called strict-feedback form [9]. It provides a systematic way of designing Lyapunov functions and nonlinear controllers for systems of any order. Unlike the feedback lineariza-tion method, backstepping can accommodate model uncertainties and avoid the unnecessary cancellation of "useful" (stabilizing) nonlinearities.

Since the dynamic model of the individual agents in this book have at most order two, we illustrate the backstepping technique by considering the system

$$\dot{x} = f(x) + \eta \tag{C.8}$$

$$\dot{\eta} = u \tag{C.9}$$

where $[x, \eta] \in \mathbb{R}^2$ is the system state, $u \in \mathbb{R}$ is the control input, and $f(x)$ is continuously differentiable with $f(0) = 0$. Say that our control objective is to stabilize the system at the equilibrium point $[x, \eta] = 0$ for any initial conditions.

Notice that the above system is a cascaded connection of subsystems (C.8) and (C.9). The idea behind backstepping is to first consider η as a control input for subsystem (C.8). Under this assumption, we could design $\eta = -f(x) - x$ to obtain the exponentially stable closed-loop system $\dot{x} = -x$. Since in reality η is a system state and thus cannot be directly manipulated, we use the trick of adding and subtracting a *fictitious* control input η_f to the right-hand side of (C.8) and introducing the variable transformation

$$\xi = \eta - \eta_f.$$

As a result, our system becomes

$$\dot{x} = f(x) + \eta_f + \xi$$
$$\dot{\xi} = u - \dot{\eta}_f.$$

Now, if we design

$$\eta_f = -f(x) - x$$
$$u = \dot{\eta}_f - \xi - x \qquad (C.10)$$

where

$$\dot{\eta}_f = \frac{\partial \eta_f}{\partial x}(f(x) + \eta),$$

we get the closed-loop system

$$\dot{x} = -x + \xi$$
$$\dot{\xi} = -\xi - x \qquad (C.11)$$

whose unique equilibrium point is $[x, \xi] = 0$.

Using the Lyapunov function candidate

$$V(x, \xi) = \frac{1}{2}(x^2 + \xi^2)$$

and taking its time derivative along (C.11) yields

$$\dot{V} = -x^2 - \xi^2.$$

From Corollary C.1, we can conclude that $[x, \xi] = 0$ is exponentially stable. Since $\eta_f(x = 0) = 0$, we know that $[x, \eta] = 0$ is an exponentially stable equilibrium point for (C.8) and (C.9) in closed-loop with (C.10).

Appendix D

Dynamic Model Terms

The matrices $M_i(\theta_i)$ and $C_i(\theta_i, \dot{\theta}_i)$ in (4.14) are defined as follows

$$M_i = \begin{bmatrix} m_i\cos^2\theta_i + \dfrac{\bar{I}_i}{L_i^2}\sin^2\theta_i & \left(m_i - \dfrac{\bar{I}_i}{L_i^2}\right)\sin\theta_i\cos\theta_i \\[2mm] \left(m_i - \dfrac{\bar{I}_i}{L_i^2}\right)\sin\theta_i\cos\theta_i & m_i\sin^2\theta_i + \dfrac{\bar{I}_i}{L_i^2}\cos^2\theta_i \end{bmatrix} \qquad \text{(D.1)}$$

and

$$C_i = \begin{bmatrix} -\left(m_i - \dfrac{\bar{I}_i}{L_i^2}\right)\dot{\theta}_i\sin\theta_i\cos\theta_i & m_i\dot{\theta}_i\cos^2\theta_i + \dfrac{\bar{I}_i}{L_i^2}\dot{\theta}_i\sin^2\theta_i \\[2mm] -m_i\dot{\theta}_i\sin^2\theta_i - \dfrac{\bar{I}_i}{L_i^2}\dot{\theta}_i\cos^2\theta_i & \left(m_i - \dfrac{\bar{I}_i}{L_i^2}\right)\dot{\theta}_i\sin\theta_i\cos\theta_i \end{bmatrix}. \qquad \text{(D.2)}$$

Notice that the eigenvalues of (D.1) are m_i and $\bar{I}_i/L_i^2$, which are both positive. Therefore, it follows from Theorem A.1 that the constants in (4.15) are given by

$$m_{i1} = \min\{m_i, \bar{I}_i/L_i^2\} \quad \text{and} \quad m_{i2} = \max\{m_i, \bar{I}_i/L_i^2\}.$$

The elements of the regression matrix Y_i in (4.17) are defined as follows

$$[Y_i]_{11} = \begin{bmatrix} \cos^2\theta_i & \sin\theta_i\cos\theta_i \end{bmatrix}\ddot{\mu} + \dot{\theta}_i\begin{bmatrix} -\sin\theta_i\cos\theta_i & \cos^2\theta_i \end{bmatrix}\dot{\mu},$$

$$[Y_i]_{12} = \begin{bmatrix} \sin^2\theta_i & -\sin\theta_i\cos\theta_i \end{bmatrix}\ddot{\mu} + \dot{\theta}_i\begin{bmatrix} \sin\theta_i\cos\theta_i & \sin^2\theta_i \end{bmatrix}\dot{\mu},$$

$$[Y_i]_{13} = \begin{bmatrix} \cos^2\theta_i & \sin\theta_i\cos\theta_i \end{bmatrix}\dot{q}_i, \qquad [Y_i]_{14} = \begin{bmatrix} -\cos\theta_i\sin\theta_i & \cos^2\theta_i \end{bmatrix}\dot{q}_i,$$

$$[Y_i]_{15} = -\begin{bmatrix} \sin\theta_i\cos\theta_i & \sin^2\theta_i \end{bmatrix}\dot{q}_i, \qquad [Y_i]_{16} = \begin{bmatrix} \sin^2\theta_i & -\cos\theta_i\sin\theta_i \end{bmatrix}\dot{q}_i,$$

$$[Y_i]_{21} = \begin{bmatrix} \sin\theta_i\cos\theta_i & \sin^2\theta_i \end{bmatrix}\ddot{\mu} + \dot{\theta}_i\begin{bmatrix} -\sin^2\theta_i & \sin\theta_i\cos\theta_i \end{bmatrix}\dot{\mu},$$

Formation Control of Multi-Agent Systems: A Graph Rigidity Approach, First Edition.
Marcio de Queiroz, Xiaoyu Cai, and Matthew Feemster.
© 2019 John Wiley & Sons Ltd. Published 2019 by John Wiley & Sons Ltd.
Companion website: www.wiley.com/go/dequeiroz/formation_control

$$[Y_i]_{22} = \left[-\sin\theta_i\cos\theta_i \quad \cos^2\theta_i\right]\dot{\mu} - \dot{\theta}_i\left[\cos^2\theta_i \quad \sin\theta_i\cos\theta_i\right]\mu,$$

$$[Y_i]_{23} = \left[\sin\theta_i\cos\theta_i \quad \sin^2\theta_i\right]\dot{q}_i, \quad [Y_i]_{24} = \left[-\sin^2\theta_i \quad \sin\theta_i\cos\theta_i\right]\dot{q}_i$$

$$[Y_i]_{25} = \left[\cos^2\theta_i \quad \sin\theta_i\cos\theta_i\right]\dot{q}_i, \quad [Y_i]_{26} = \left[-\sin\theta_i\cos\theta_i \quad \cos^2\theta\right]\dot{q}_i.$$

References

1 T. Holbrook, R. Clark, and B. Haney, Secrets of a superorganism, *ASU – Ask a Biologist*, Sept. 2009, http://askabiologist.asu.edu/explore/secrets-superorganism.

2 V. Gazi and K.M. Passino, *Swarm stability and optimization*, Berlin: Springer-Verlag, 2011.

3 M. Mesbahi and M. Egerstedt, *Graph theoretic methods in multiagent networks*, Princeton, NJ: Princeton University Press, 2010.

4 W. Ren and Y. Cao, *Distributed coordination of multi-agent networks: emergent problems, models, and issues*, London: Springer-Verlag, 2011.

5 S. Martinez and F. Bullo, Optimal sensor placement and motion coordination for target tracking, *Automatica*, vol. 42, no. 3. pp. 661–668, 2006.

6 B.D.O. Anderson, C. Yu, B. Fidan, and J.M. Hendrickx, Rigid graph control architectures for autonomous formations, *IEEE Contr. Syst. Mag.*, vol. 28, no. 6, pp. 48–63, 2008.

7 C. Yu, B.D.O. Anderson, S. Dasgupta, and B. Fidan, Control of minimally persistent formations in the plane, *SIAM J. Contr. Optim.*, vol. 48, no. 1, pp. 206–233, 2009.

8 H.K. Khalil, *Nonlinear systems*, Englewood Cliffs, NJ: Prentice Hall, 2002.

9 M. Krstic, I. Kanellakopoulos, and P. Kokotovic, *Nonlinear and adaptive control design*, New York, NY: John Wiley & Sons, 1995.

10 Y. Zhao, Z. Duan, G. Wen, and Y. Zhang, Distributed finite-time tracking control for multi-agent systems: An observer-based approach, *Syst. & Contr. Lett.*, vol. 62, pp. 22–28, 2013.

11 F.P. Beer, E.R. Johnston, and W.E. Clausen, *Vector mechanics for engineers: Dynamics*, New York, NY: McGraw Hill, 2007.

12 L. Euler, *Opera postuma*, vol. 1, pp. 494–496, 1862. (See: The Euler Archive, index number 819, http://eulerarchive.maa.org.)

13 B. Jackson, Notes on the rigidity of graphs, *Notes of the Levico Conference*, Levico Terme, Italy, 2007.

Formation Control of Multi-Agent Systems: A Graph Rigidity Approach, First Edition.
Marcio de Queiroz, Xiaoyu Cai, and Matthew Feemster.
© 2019 John Wiley & Sons Ltd. Published 2019 by John Wiley & Sons Ltd.
Companion website: www.wiley.com/go/dequeiroz/formation_control

14 I. Izmestiev, Infinitesimal rigidity of frameworks and surfaces, *Lectures on Infinitesimal Rigidity*, Kyushu University, Japan, 2009.

15 L. Asimow and B. Roth, The rigidity of graphs II, *J. Math. Anal. Appl.*, vol. 68, no. 1, pp. 171–190, 1979.

16 B. Servatius and H. Servatius, Generic and abstract rigidity, in *Rigidity Theory and Applications*, pp. 1–19, New York, NY: Springer-Verlag, 2002.

17 J. Graver, B. Servatius, and H. Servatius, *Combinatorial rigidity*, Providence, RI: American Mathematical Society, 1993.

18 B. Roth, Rigid and flexible frameworks, *The Amer. Math. Monthly*, vol. 86, no. 1, pp. 6–21, 1981.

19 J.N. Franklin, *Matrix theory*, Englewood Cliffs, NJ: Prentice Hall, 1968.

20 B. Hendrickson, Conditions for unique graph realizations, *SIAM J. Comput.*, vol. 21, no.1, pp. 65–84, 1992.

21 D. Moore, J. Leonard, D. Rus, and S. Teller, Robust distributed network localization with noisy range measurements, *Proc. ACM Conf. Embedded Networked Sensor Syst.*, pp. 50–61, 2004.

22 J. Aspnes, J. Egen, D.K. Goldenberg, A.S. Morse, W. Whiteley, Y.R. Yang, B.D.O. Anderson, and P.N. Belhumeur, A theory of network localization, *IEEE Trans. Mob. Comput.*, vol. 5, no. 12, pp. 1663–1678, 2006.

23 L. Krick, M.E. Broucke, and B.A. Francis, Stabilization of infinitesimally rigid formations of multi-robot networks, *Intl. J. Contr.*, vol. 83, no. 3, pp. 423–439, 2009.

24 B. Xian, D.M. Dawson, M. de Queiroz, and J. Chen, A continuous asymptotic tracking control strategy for uncertain nonlinear systems, *IEEE Trans. Autom. Contr.*, vol. 49, no.7, pp. 1206–1211, 2004.

25 L. Asimow and B. Roth, The rigidity of graphs, *Trans. Amer. Math. Soc.*, vol. 245, pp. 279–289, 1978.

26 R. Connelly, Generic global rigidity, *Discrete Comput. Geom.*, vol. 33, no.4, pp. 549–563, 2005.

27 R. Diestel, *Graph theory*, New York: Springer-Verlag, 1997.

28 B. Jackson and T. Jordán, Connected rigidity matroids and unique realizations of graphs, *J. Comb. Theory, Series B*, vol. 94, no. 1, pp. 1–29, 2005.

29 H. Maehara, Geometry of frameworks, *Yokohama Math. J.*, vol. 47, pp. 41–65, 1999.

30 T. Eren, P.N. Belhumeur, and A.S. Morse, Closing ranks in vehicle formations based on rigidity, *Proc. IEEE Conf. Dec. Contr.*, pp. 2959–2964, Las Vegas, NV, 2002.

31 J. Baillieul and A. Suri, Information patterns and hedging Brockett's theorem in controlling vehicle formations, *Proc. IEEE Conf. Dec. Contr.*, pp. 556–563, Maui, HI, 2003.

32 J.P. Desai, J. Ostrowski, and V. Kumar, Controlling formations of multiple mobile robots, *Proc. IEEE Intl. Conf. Rob. Autom.*, pp. 2864–2869, Leuven, Belgium, 1998.

33 J.A. Fax and R.M. Murray, Information flow and cooperative control of vehicle formations, *IEEE Trans. Autom. Contr.*, vol. 49, no. 9, pp. 1465–1476, 2004.

34 R. Olfati-Saber and R.M. Murray, Distributed cooperative control of multiple vehicle formations using structural potential functions, Presented at the *15th IFAC World Congress*, Barcelona, Spain, 2002.

35 P. Tabuada, G.J. Pappas, and P. Lima, Feasible formations of multi-agent systems, *Proc. Amer. Contr. Conf.*, pp. 56–61, Arlington, VA, 2001.

36 V. Gazi, B. Fidan, R. Ordóñez, and M.I. Köksal, A target tracking approach for nonholonomic agents based on artificial potentials and sliding mode control, *ASME J. Dyn. Syst., Meas., and Contr.*, vol. 134, no. 6, Paper 061004, 2012.

37 H. Bai, M. Arcak, and J. Wen, *Cooperative control design: A systematic, passivity-based approach*, New York, NY: Springer, 2010.

38 Z. Qu, *Cooperative control of dynamical systems: applications to autonomous vehicles*, London: Springer-Verlag, 2009.

39 W. Ren and R.W. Beard, *Distributed consensus in multi-vehicle cooperative control*, London: Springer-Verlag, 2008.

40 K.-K. Oh, M.-C. Park, and H.-S. Ahn, A survey of multi-agent formation control, *Automatica*, vol. 53, no. 3, pp. 424–440, 2015.

41 G. Campion, G. Bastin, and B. D'Andrea-Novel, Structural properties and classification of kinematic and dynamic models of wheeled mobile robots, *IEEE Trans. Rob. Autom.*, vol. 12, no.1, pp. 47–62, 1996.

42 D. Shevitz and B. Paden, Lyapunov stability of nonsmooth systems, *IEEE Trans. Autom. Contr.*, vol. 39, no.9, pp. 1910–1914, 1994.

43 Z. Qu and J.X. Xu, Model-based learning controls and their comparisons using Lyapunov direct method, *Asian J. Contr.*, vol. 4, no. 1, pp. 99–110, 2002.

44 R. Kamalapurkar, J. Klotz, R. Downey, and W.E. Dixon, Supporting lemmas for RISE-based control methods, arXiv:1306.3432 [cs.SY], 2013.

45 A. Ben-Israel and T.N.E. Greville, *Generalized inverses: Theory and applications*, New York, NY: Springer-Verlag, 2003.

46 D.V. Dimarogonas and K.H. Johansson, On the stability of distance-based formation control, *Proc. Conf. Dec. Contr.*, pp. 1200–1205, Cancun, Mexico, 2008.

47 K.-K. Oh and H.-S. Ahn, Formation control of mobile agents based on inter-agent distance dynamics, *Automatica*, vol. 47, no. 10, pp. 2306–2312, 2011.

48 K.-K. Oh and H.-S. Ahn, Distance-based undirected formations of single-integrator and double-integrator modeled agents in n-dimensional space, *Intl. J. Rob. Nonl. Contr.*, vol. 24, no. 12, pp. 1809–1820, 2014.

49 Z. Sun, S. Moub, B.D.O. Anderson, and M. Cao, Exponential stability for formation control systems with generalized controllers: A unified approach, *Syst. Contr. Lett.*, vol. 93, no. 7, pp. 50–57, 2016.

50 S.-M Kang, M.-C. Park, B.-H. Lee, and H.-S. Ahn, Distance-based formation control with a single moving leader, *Proc. Amer. Contr. Conf.*, pp. 305–310, Portland, OR, 2014.

51 K.-K. Oh and H.-S. Ahn, Distance-based control of cycle-free persistent formations, *Proc. IEEE Multi-Conf. Syst. Contr.*, pp. 816–821, Denver, CO, 2011.

52 O. Rozenheck, S. Zhao, and D. Zelazo, A proportional-integral controller for distance-based formation tracking, *Proc. Eur. Contr. Conf.*, pp. 1693–1698, Linz, Austria, 2015.

53 Z. Sun, M.-C. Park, B.D.O. Anderson, and H.-S. Ahn, Distributed stabilization control of rigid formations with prescribed orientation, *Automatica*, vol. 78, no. 4, pp. 250–257, 2017.

54 B.D.O. Anderson, Z. Sun, T. Sugie, S. Azuma, and K. Sakurama, Formation shape control with distance and area constraints, *IFAC J. Syst. and Contr.*, vol. 1, pp. 2–12, 2017.

55 F. Dörfler and B. Francis, Geometric analysis of the formation problem for autonomous robots, *IEEE Trans. Autom. Contr.*, vol. 55, no. 10, pp. 2379–2384, 2010.

56 K.-K. Oh and H.-S. Ahn, Formation control of mobile agents based on distributed position estimation, *IEEE Trans. Autom. Contr.*, vol. 58, no. 3, pp. 737–742, 2013.

57 T.H. Summers, C. Yu, S. Dasgupta, and B.D.O. Anderson, Control of minimally persistent leader-remote-follower and coleader formations in the plane, *IEEE Trans. Autom. Contr.*, vol. 56, no. 12, pp. 2778–2792, 2011.

58 S. Mou, M.-A. Belabbas, A.S. Morse, Z. Sun, and B.D.O. Anderson, Undirected rigid formations are problematic, *IEEE Trans. Autom. Contr.*, vol. 61, no. 10, pp. 2821–2836, 2016.

59 Z. Sun, S. Mou, B.D.O. Anderson, and A.S. Morse, Rigid motions of 3-D undirected formations with mismatch between desired distances, *IEEE Trans. Autom. Contr.*, vol. 62, no. 8, pp. 4151–4158, 2017.

60 H.G. de Marina, M. Cao, and B. Jayawardhana, Controlling rigid formations of mobile agents under inconsistent measurements, *IEEE Trans. Rob.*, vol. 31, no. 1, pp. 31–39, 2015.

61 H.G. Tanner, G.J. Pappas, and V. Kumar, Input-to-state stability on formation graphs, *Proc. Conf. Dec. Control*, pp. 2439–2444, Las Vegas, NV, 2002.

62 T. Eren, W. Whiteley, B.D.O. Anderson, A.S. Morse, and P.N. Bellhumeur, Information structures to secure control of rigid formations with leader-follower structure, *Proc. Amer. Contr. Conf.*, pp. 2966–2971, Portland, OR, 2005.

63 J.M. Hendrickx, B.D.O. Anderson, J.-C. Delvenne, and V.D. Blondel, Directed graphs for the analysis of rigidity and persistence in autonomous agent systems, *Intl. J. Rob. Nonl. Contr.*, vol. 17, no. 10, pp. 960–981, 2007.

64 C. Yu, J.M. Hendrickx, B. Fidan, B.D.O. Anderson, and V.D. Blondel, Three and higher dimensional autonomous formations: Rigidity, persistence and structural persistence, *Automatica*, vol. 43, pp. 387–402, 2007.

65 B.D.O. Anderson, C. Yu, S. Dasgupta, and A.S. Morse, Control of a three-coleader formation in the plane, *Syst. Contr. Lett.*, vol. 56, pp. 573–578, 2007.

66 M. Cao, A.S. Morse, C. Yu, B.D.O. Anderson, and S. Dasgupta, Maintaining a directed, triangular formation of mobile autonomous agents, *Commun. Inf. Syst.*, vol. 11, no. 1, pp. 1–16, 2011.

67 M.-C. Park, Z. Sun, K.-K. Oh, B.D.O. Anderson, and H.-S. Ahn, Finite-time convergence control for acyclic persistent formations, *IEEE Intl. Symp. Intell. Contr.*, pp. 1608–1613, Antibes, France, 2014.

68 F. Xiao, L. Wang, J. Chen, and Y. Gao, Finite-time formation control for multi-agent systems, *Automatica*, vol. 45, no. 11, pp. 2605–2611, 2009.

69 Z. Lin, L. Wang, Z. Han, and M. Fu, Distributed formation control of multi-agent systems using complex Laplacian, *IEEE Trans. Autom. Contr.*, vol. 59, no. 7, pp. 1765–1777, 2014.

70 M. Basiri, A.N. Bishop, and P. Jensfelt, Distributed control of triangular formations with angle-only constraints, *Syst. Contr. Lett.*, vol. 59, pp. 147–154, 2010.

71 L. Wang, J. Markdahl, and X. Hu, Distributed attitude control of multi-agent formations, *Proc. IFAC World Congr.*, pp. 4513–4518, Milano, Italy, 2011.

72 H.G. de Marina, B. Jayawardhana, and M. Cao, Distributed rotational and translational maneuvering of rigid formations and their applications, *IEEE Trans. Rob.*, vol. 32, no. 3, pp. 684–697, 2016.

73 X. Cai and M. de Queiroz, On the stabilization of planar multi-agent formations, *Proc. ASME Conf. Dyn. Syst. Contr.*, paper no. DSCC2012-MOVIC2012-8534, Ft. Lauderdale, FL, 2012.

74 X. Cai and M. de Queiroz, Formation maneuvering and target interception for multi-agent systems via rigid graphs, *Asian J. Contr.*, vol. 17, no. 4, pp. 1174–1186, 2015.

75 P. Zhang, M. de Queiroz, and X. Cai, 3D dynamic formation control of multi-agent systems using rigid graphs, *ASME J. Dyn. Syst. Measur. Contr.*, vol. 137, no. 11, Paper no. 111006, 2015.

76 C. Cao and W. Ren, Distributed coordinated tracking with reduced interaction via a variable structure approach, *IEEE Trans. Autom. Contr.*, vol. 57, no. 1, pp 33–48, 2012.

77 J. Mei, W. Ren, and G. Ma, Distributed coordinated tracking with a dynamic leader for multiple Euler-Lagrange systems, *IEEE Trans. Autom. Contr.*, vol. 56, no. 6, pp. 1415–1421, 2011.

78 Z. Sun, B.D.O. Anderson, M. Deghat, and H.-S. Ahn, Rigid formation control of double-integrator systems, *Intl. J. Contr.*, vol. 90, no. 7, pp. 1403–1419, 2017.

79 P. Ögren, E. Fiorelli, and N.E. Leonard, Cooperative control of mobile sensor networks: Adaptive gradient climbing in a distributed environment, *IEEE Trans. Autom. Contr.*, vol. 49, no. 8, pp. 1292–1302, 2004.

80 X. Dong, B. Yu, Z. Shi, and Y. Zhong, Time-varying formation control for unmanned aerial vehicles: Theories and applications, *IEEE Trans. Contr. Syst. Tech.*, vol. 23, no. 1, pp. 340–348, 2015.

81 H. Bai, M. Arcak, and J. T. Wen, Using orientation agreement to achieve planar rigid formation, *Proc. Amer. Contr. Conf.*, pp. 753–758, Seattle, WA, 2008.

82 D. Sun, C. Wang, W. Shang, and G. Feng, A synchronization approach to trajectory tracking of multiple mobile robots while maintaining time-varying formations, *IEEE Trans. Rob.*, vol. 25, no. 5, pp. 1074–1086, 2009.

83 Z. Han, L. Wang, Z. Lin, and R. Zheng, Formation control with size scaling via a complex Laplacian-based approach, *IEEE Trans. Cybern.*. vol. 46, no. 1, pp. 2348–2359, 2016.

84 Z. Sun, S. Mou, M. Deghat, and B.D.O. Anderson, Finite time distributed distance-constrained shape stabilization and flocking control for d-dimensional undirected rigid formations, *Intl. J. Rob. Nonl. Contr.*, vol. 26, no. 13, pp. 2824–2844, 2016.

85 M. Deghat, B.D.O. Anderson, and Z. Lin, Combined flocking and distance-based shape control of multi-agent formations, *IEEE Trans. Autom. Contr.*, vol. 61, no. 7, pp. 1824–1837, 2016.

86 Y. Cao, D. Stuart, W. Ren, and Z. Meng, Distributed containment control for multiple autonomous vehicles with double-integrator dynamics: algorithms and experiments, *IEEE Trans. Contr. Syst. Tech.*, vol. 19, no. 4, pp. 929–938, 2011.

87 Y.-Y. Chen and Y.-P. Tian, A backstepping design for directed formation control of three-coleader agents in the plane, *Intl. J. Rob. Nonl. Contr.*, vol. 19, no. 7, pp. 729–745, 2009.

88 S. Coogan and M. Arcak, Scaling the size of a formation using relative position feedback, *Automatica*, vol. 48, no. 10, pp. 2677–2685, 2012.

89 H.G. de Marina, B. Jayawardhana, and M. Cao, Taming mismatches in inter-agent distances for the formation-motion control of second-order agents, *IEEE Trans. Autom. Contr.*, vol. 63, no. 2, pp. 449–462, 2018.

90 X. Cai and M. de Queiroz, Multi-agent formation maneuvering and target interception with double-integrator model, *Proc. Amer. Contr. Conf.*, pp. 287–292, Portland, OR, 2014.

91 X. Cai and M. de Queiroz, Rigidity-based stabilization of multi-agent formations, *ASME J. Dyn. Syst. Measur. Contr.*, vol. 136, no. 1, Paper 014502, 2014.

92 W.E. Dixon, D.M. Dawson, E. Zergeroglu, and A. Behal, *Nonlinear control of wheeled mobile robots*, London: Springer, 2001.

93 Y. Fang, E. Zergeroglu, M.S. de Queiroz, and D.M. Dawson, Global output feedback control of dynamically positioned surface vessels: An adaptive control approach, *Mechatronics*, vol. 14, no. 4, pp. 341–356, 2004.

94 A. De Luca and G. Oriolo, Modelling and control of nonholonomic mechanical systems, in J. Angeles, A. Kecskemethy (eds.) *Kinematics and Dynamics of Multi-Body Systems*, CISM Courses and Lectures, Vol. 360, pp. 277–342, Wien, Germany: Springer-Verlag, 1995.

95 R.W. Brockett, Asymptotic stability and feedback stabilization, in R.W. Brockett, R.S. Millman, and H.J. Sussmann (eds.) *Differential Geometric Control Theory*, pp. 181–191, Boston, MA: Birkhauser, 1983.

96 S. Mastellone, D.M. Stipanovic, C.R. Graunke, K.A. Intlekofer, and M.W. Spong, Formation control and collision avoidance for multi-agent non-holonomic systems: Theory and experiments, *Intl. J. Robotics Res.*, vol. 27, no. 1, pp. 107–125, 2008.

97 D. Kostic, S. Adinandra, J. Caarls, N. van de Wouw, and H. Nijmeijer, Saturated control of time-varying formations and trajectory tracking for unicycle multi-agent systems, *Proc. IEEE Conf. Dec. Contr.*, pp. 4054–4059, Atlanta, GA, 2010.

98 A. Sadowska, D. Kostic, N. van de Wouw, H. Huijberts, and H. Nijmeijer, Distributed formation control of unicycle robots, *Proc. IEEE Intl. Conf. Rob. Autom.*, pp. 1564–1569, Saint Paul, MN, 2012.

99 N. Moshtagh, N. Michael, A. Jadbabaie, and K. Daniilidis, Vision-based, distributed control laws for motion coordination of nonholonomic robots, *IEEE Trans. Rob.*, vol. 25, no. 4, pp. 851–860, 2009.

100 M. Khaledyan and M. de Queiroz, Formation maneuvering control of non-holonomic multi-agent systems, *Proc. ASME Dyn. Syst. Contr. Conf.*, Paper No. DSCC2016-9616, Minneapolis, MN, 2016.

101 B.-H. Lee, S.-J. Lee, M.-C. Park, K.-K. Oh, and H.-S. Ahn, Nonholonomic control of distance-based cyclic polygon formation, *Proc. Asian Contr. Conf.*, pp. 1–4, Istanbul, Turkey, 2013.

102 D.V. Dimarogonas and K.J. Kyriakopoulos, On the rendevous problem for mulitple nonholonomic agents, *IEEE Trans. Autom. Contr.*, vol. 52, no. 5, pp. 916–922, 2007.

103 D.V. Dimarogonas and K.J. Kyriakopoulos, A connection between formation infeasibility and velocity alignment in kinematic multi-agent systems, *Automatica*, vol. 44, no. 10, pp. 2648–2654, 2009.

104 Y. Liang and H.-H Lee, Decentralized formation control and obstacle avoidance for multiple robots with nonholonomic constraints, *Proc. Amer. Contr. Conf.*, pp. 5596–5601, Minneapolis, MN, 2006.

105 J. Zhu, J. Lü, and X. Yu, Flocking of multi-agent non-holonomic systems with proximity graphs, *IEEE Trans. Circ. Syst. I*, vol. 60, no. 1, pp. 199–210, 2013.

106 J. Chen, D. Sun, J. Yang, and H. Chen, Leader-follower formation control of multiple non-holonomic mobile robots incorporating a receding-horizon scheme, *Intl. J. Rob. Res.*, vol. 29, no. 6, pp. 727–747, 2010.

107 S.-J. Chung and J.-J. Slotine, Cooperative robot control and concurrent synchronization of Lagrangian systems, *IEEE Trans. Rob.*, vol. 25, no. 3, pp. 686–700, 2009.

108 S. Khoo, L. Xie, and Z. Man, Robust finite-time consensus tracking algorithm for multirobot systems, *IEEE/ASME Trans. Mechatr.*, vol. 14. no. 2, pp. 219–228, 2009.

109 A.R. Pereira, L. Hsu, and R. Ortega, Globally stable adaptive formation control of Euler–Lagrange agents via potential functions, *Proc. Amer. Contr. Conf.*, pp. 2606–2611, St. Louis, MO, 2009.

110 G. Chen and F.L. Lewis, Distributed adaptive tracking control for synchronization of unknown networked Lagrangian systems, *IEEE Trans. Syst. Man Cybern. – Part B*, vol. 41, no. 3, pp. 805–816, 2011.

111 D. Lee and P.Y. Li, Passive decomposition approach to formation and maneuver control of multiple rigid-bodies, *ASME J. Dyn. Syst. Measur. Contr.*, vol. 129, no. 5, pp. 662–677, 2007.

112 Z. Peng, D. Wang, Z. Chen, X. Hu, and W. Lan, Adaptive dynamic surface control for formations of autonomous surface vehicles with uncertain dynamics, *IEEE Trans. Contr. Syst. Tech.*, vol. 21, no. 2, pp. 513–520, 2013.

113 R. Skjetne, S. Moi, and T.I. Fossen, Nonlinear formation control of marine craft, *Proc. IEEE Conf. Dec. Contr.*, pp. 1699–1704, Las Vegas, NV, 2002.

114 T. Dierks and S. Jagannathan, Control of nonholonomic mobile robot formations: backstepping kinematics into dynamics, *Proc. Intl. Conf. Contr. Appl.*, pp. 94–99, Singapore, 2007.

115 K.D. Do and J. Pan, Nonlinear formation control of unicycle-type mobile robots, *Robot. Autonom. Syst.*, vol. 55, pp. 191–204, 2007.

116 W. Dong and J.A. Farrell, Decentralized cooperative control of multiple nonholonomic dynamic systems with uncertainty, *Automatica*, vol. 45, no. 3, pp. 706–710, 2009.

117 C.F.L. Thorvaldsen and R. Skjetne, Formation control of fully-actuated marine vessels using group agreement protocols, *Proc. IEEE Conf. Dec. Contr.*, pp. 4132–4139, Orlando, FL, 2011.

118 J. Yao, R. Ordóñez, and V. Gazi, Swarm tracking using artificial potentials and sliding mode control, *ASME J. Dyn. Syst., Meas., Control*, vol. 129, no. 5, pp. 749–754, 2007.

119 X. Cai and M. de Queiroz, Adaptive rigidity-based formation control for multi-robotic vehicles with dynamics, *IEEE Trans. Contr. Syst. Tech.*, vol. 23, no. 1, pp. 389–396, 2015.

120 M. Khaledyan and M. de Queiroz, Translational maneuvering control of nonholonomic kinematic formations: Theory and experiments, *Proc. Amer. Contr. Conf.*, pp. 2910–2915, Milwaukee, WI, 2018.

121 A. De Luca, G. Oriolo, and C. Samson, Feedback control of a nonholonomic car-like robot, in J.-P. Laumond (ed.), *Robot Motion Planning and Control*, Lectures Notes in Control and Information Sciences, Vol. 229, pp. 171–253, Berlin: Springer-Verlag, 1998.

122 E. Moret, *Dynamic modeling and control of a car-like robot*, Master thesis, Virginia Polytechnic Institute, 2003.

123 Z. Cai, M.S. de Queiroz, and D.M. Dawson, A Sufficiently Smooth Projection Operator,textquotedblright *IEEE Trans. Autom. Contr.*, vol. 51, no. 1, pp. 135–139, Jan. 2006.

124 K.S. Narendra and A.M. Annaswamy, *Stable adaptive systems*. Mineola, NY: Dover, 2005.

125 P.R. Halmos, *Finite-dimensional vector spaces*, New York, NY: Springer-Verlag, 1974.

126 R.A. Horn and C.R. Johnson, *Matrix analysis*, Cambridge, UK: Cambridge University Press, 1985.

127 R. Penrose, A generalized inverse for matrices, *Proc. Cambridge Phil. Soc.*, vol. 51, pp. 406–413, 1955.

128 D.M. Dawson, J. Hu, and T.C. Burg, *Nonlinear control of electric machinery*, New York, NY: Marcel Dekker, 1998.

129 C.A. Desoer and M. Vidyasagar, *Feedback systems: Input-output properties*, New York, NY: Academic Press, 1975.

130 F. Golnaraghi and B.C. Kuo, *Automatic control systems*, New York, NY: McGraw-Hill Education, 2017.

131 S. Sastry, *Nonlinear systems: analysis, stability, and control*, New York, NY: Springer, 1999.

132 J.-J.E. Slotine and W. Li, *Applied nonlinear control*, Englewood Cliffs, NJ: Prentice Hall, 1991.

133 Khalil, H. K., *Nonlinear control*, Harlow, England: Pearson Education Limited, 2015.

134 J.H. Marquez, Nonlinear control systems analysis and design, Hoboken, NY: John Wiley & Sons, 2003.

135 A.F. Filippov, *Differential equations with discontinuous righthand sides*, Dordrecht, The Netherlands: Kluwer Academic Publishers, 2010.

136 Z. Guo and L. Huang, Generalized Lyapunov method for discontinuous systems, *Nonl. Anal.*, vol. 71, pp. 3083–3092, 2009.

Index

Formation Control of Multi-Agent Systems: A Graph Rigidity Approach, First Edition.
Marcio de Queiroz, Xiaoyu Cai, and Matthew Feemster.
© 2019 John Wiley & Sons Ltd. Published 2019 by John Wiley & Sons Ltd.
Companion website: www.wiley.com/go/dequeiroz/formation_control